SpringerBriefs in Mathematical Physics

Volume 32

SpringerBriefs are characterized in general by their size (50–125 pages) and fast production time (2–3 months compared to 6 months for a monograph).

Briefs are available in print but are intended as a primarily electronic publication to be included in Springer's e-book package.

Typical works might include:

- An extended survey of a field
- A link between new research papers published in journal articles
- A presentation of core concepts that doctoral students must understand in order to make independent contributions
- Lecture notes making a specialist topic accessible for non-specialist readers.

SpringerBriefs in Mathematical Physics showcase, in a compact format, topics of current relevance in the field of mathematical physics. Published titles will encompass all areas of theoretical and mathematical physics. This series is intended for mathematicians, physicists, and other scientists, as well as doctoral students in related areas.

More information about this series at http://www.springer.com/series/11953

Karl-Hermann Neeb · Gestur Ólafsson

Reflection Positivity

A Representation Theoretic Perspective

 Springer

Karl-Hermann Neeb
Department Mathematik
Universität Erlangen-Nürnberg
Erlangen, Germany

Gestur Ólafsson
Department of Mathematics
Louisiana State University
Baton Rouge, LA, USA

ISSN 2197-1757   ISSN 2197-1765   (electronic)
SpringerBriefs in Mathematical Physics
ISBN 978-3-319-94754-9   ISBN 978-3-319-94755-6   (eBook)
https://doi.org/10.1007/978-3-319-94755-6

Library of Congress Control Number: 2018946582

Printed on acid-free paper

This Springer imprint is published by the registered company Springer International Publishing AG part of Springer Nature
The registered company address is: Gewerbestrasse 11, 6330 Cham, Switzerland

Preface

The concept of reflection positivity (RP) occurs as an important theme in various areas of mathematics and physics:

- In the representation theory of Lie groups, it establishes a passage of a unitary representation of a symmetric Lie group (such as the euclidean motion group) to a unitary representation of the Cartan dual group (such as the Poincaré group) [FOS83, JO100, JO198, NÓ14, JPT15].
- In constructive Quantum Field Theorem (QFT), it arises as the condition of Osterwalder–Schrader (OS) positivity for a euclidean field theory to correspond to a relativistic one [GJ81, Ja08, Ja18, Os95a, Os95b, OS73, OS75].
- For stochastic processes, it is weaker than the Markov property and specifies processes arising in lattice gauge theory. It plays a central role in the mathematical study of phase transitions and symmetry breaking [FILS78, JJ16, JJ17, Nel73].
- In analysis, it is a crucial condition that leads to inequalities such as the Hardy–Littlewood–Sobolev inequality [FL10].

Only recently, it became apparent that there are many hidden and still not sufficiently well-understood structures underlying the duality between unitary representations of a symmetric Lie group and its dual. Establishing reflection positivity in this context requires new analytic methods and new geometric insight into constructions and realizations of representations in analytic contexts. New developments concern analytic issues such as criteria for integrating Lie algebra representations to Lie group representations, reflection positive functions, distributions and kernels, new dilation techniques for representations and unexpected connections between Kubo–Martin–Schwinger (KMS) states of C^*-algebras and reflection positive unitary representations.

This was our motivation to write this "light" introduction to the representation theoretic aspects of reflection positivity to present this perspective on a level suitable for doctoral students.

Acknowledgment: We are most grateful to Jan Frahm for reading earlier versions of this manuscript and for helping us to eliminate typos and inaccuracies.

Erlangen, Germany   Karl-Hermann Neeb
Baton Rouge, USA   Gestur Ólafsson

References

[FL10]   Frank, R.L., Lieb, E.H.: Inversion positivity and the sharp Hardy-Littlewood-Sobolev inequality. Calc. Var. Partial. Differ. Equ. **39**, 85–99 (2010)

[FILS78]   Fröhlich, J., Israel, R., Lieb, E.H., Simon, B.: Phase transitions and reflection positivity. I. General theory and long range lattice models. Commun. Math. Phys. **62**(1), 1–34 (1978)

[FOS83]   Fröhlich, J., Osterwalder, K., Seiler, E.: On virtual representations of symmetric spaces and their analytic continuation. Ann. Math. **118**, 461–489 (1983)

[GJ81]   Glimm, J., Jaffe, A.: Quantum Physics-A Functional Integral Point of View. Springer, New York (1981)

[Ja08]   Jaffe, A.: Quantum theory and relativity. In: Doran, R.S., Moore, C.C., Zimmer, R.J. (eds.) Group Representations, Ergodic Theory, and Mathematical Physics: A Tribute to George W. Mackey. Contemporary in Mathematics, vol. 449. American Mathematical Society, Providence (2008)

[Ja18]   Jaffe, A.: Reflection positivity then and now, Oberwolfach Reports 55/2017, "Reflection Positivity". doi: 10.4171/OWR/2017/55

[JJ16]   Jaffe, A., Janssens, B.: Characterization of reflection positivity: Majoranas and spins. Commun. Math. Phys. **346**(3), 1021–1050 (2016)

[JJ17]   Jaffe, A., Janssens, B.: Reflection positive doubles. J. Funct. Anal. **272**(8), 3506–3557 (2017)

[JOl00]   Jorgensen, P.E.T., Ólafsson, G.: Unitary representations and Osterwalder–Schrader duality. In: Doran R.S., Varadarajan V.S. (eds.) The Mathematical Legacy of Harish–Chandra. Proceedings of Symposia in Pure Mathematics, vol. 68. American Mathematical Society, pp 333–401, (2000)

[JOl98]   Jorgensen, P.E.T., Ólafsson, G.: Unitary representations of Lie groups with reflection symmetry. J. Funct. Anal. **158**, 26–88 (1998)

[JPT15]   Jorgensen, P.E.T., Pedersen, S., Tian, F.: Extensions of Positive Definite Functions: Applications and Their Harmonic Analysis. Lecture Notes in Mathematics, vol. 2160. Springer, Berlin (2016)

[Nel73]   Nelson, E.: Construction of quantum fields from Markoff fields. J. Funct. Anal. **12**, 97–112 (1973)

[NÓ14]   Neeb, K.-H., Ólafsson, G.: Reflection positivity and conformal symmetry. J. Funct. Anal. **266**, 2174–2224 (2014)

[Os95a]   Osterwalder, K.: Supersymmetric quantum field theory. In: "Constructive Physics," (Palaiseau, 1994), Lecture Notes in Physics 446, V. Rivasseau, Ed., pp 117–130, Springer, Berlin, (1995)

[Os95b]   Osterwalder, K., Constructing supersymmetric quantum field theory models. In: "Advances in Dynamical Systems and Quantum Physics," (Capri, 1993), pp 209–224. World Scientific Publication, River Edge, NJ (1995)

[OS73]   Osterwalder, K., Schrader, R.: Axioms for Euclidean green's functions. I. Commun. Math. Phys. **31**, 83–112 (1973)

[OS75]   Osterwalder, K., Schrader, R.: Axioms for Euclidean Green's functions. II. Commun. Math. Phys. **42**, 281–305 (1975)

Contents

1   **Introduction**   1

2   **Reflection Positive Hilbert Spaces**   9
    2.1   Reflection Positive Hilbert Spaces   9
    2.2   Reflection Positive Subspaces as Graphs.   10
    2.3   The Markov Condition   11
    2.4   Reflection Positive Kernels and Distributions   12
    2.5   Reflection Positivity in Riemannian Geometry.   16
    2.6   Selfadjoint Extensions and Reflection Positivity.   17

3   **Reflection Positive Representations**   21
    3.1   The OS Transform of Linear Operators.   21
    3.2   Symmetric Lie Groups and Semigroups   24
    3.3   Reflection Positive Representations   26
    3.4   Reflection Positive Functions   29

4   **Reflection Positivity on the Real Line**   35
    4.1   Reflection Positive Functions on Intervals.   36
    4.2   Reflection Positive One-Parameter Groups   38
    4.3   Reflection Positive Operator-Valued Functions   41
    4.4   A Connection to Lax–Phillips Scattering Theory   47

5   **Reflection Positivity on the Circle**   51
    5.1   Positive Definite Functions Satisfying KMS Conditions   52
    5.2   Reflection Positive Functions and KMS Conditions   57
    5.3   Realization by Resolvents of the Laplacian   62

6   **Integration of Lie Algebra Representations**   69
    6.1   A Geometric Version of Fröhlich's Selfadjointness Theorem   70
    6.2   Integrability for Reproducing Kernel Spaces   71
    6.3   Representations on Spaces of Distributions   74
    6.4   Reflection Positive Distributions and Representations   76

7   Reflection Positive Distribution Vectors   79
   7.1   Distribution Vectors   79
       7.1.1   Distributions on Lie Groups and Homogeneous
             Spaces   80
       7.1.2   Smooth Vectors and Distribution Vectors   81
   7.2   Reflection Positive Distribution Vectors   85
   7.3   Spherical Representation of the Lorentz Group   86
       7.3.1   The Principal Series   86
       7.3.2   The Complementary Series   88
       7.3.3 H-Invariant Distribution Vectors   92
   7.4   Reflection Positivity   93
       7.4.1   Reflection Positivity for the Conformal Group   93
       7.4.2   Resolvents of the Laplacian on the Sphere   95
       7.4.3   The Distribution Kernel of $(m^2 - \Delta)^{-1}$ 97

8   Generalized Free Fields   103
   8.1   Lorentz Invariant Measures on the Light Cone and Their
      Relatives   103
   8.2   From the Poincaré Group to the Euclidean Group   105
   8.3   The Conformally Invariant Case   110

9   Reflection Positivity and Stochastic Processes   113
   9.1   Reflection Positive Group Actions on Measure Spaces   113
   9.2   Stochastic Processes Indexed by Lie Groups   117
   9.3   Associated Positive Semigroup Structures and Reconstruction . . .   118

Appendix: Background Material   123

References   131

Index   137

Chapter 1
Introduction

In the context of quantum physics reflection positivity is often related to Wick rotation, which roughly means multiplying the time coordinate by $i = \sqrt{-1}$. This can be made precise and used in analytic constructions if this process is related to analytic continuation of the time variable to a domain in the complex plane which provides a means to go back and forth between real and imaginary time.

A duality of a similar flavor also exists in the context of Lie groups, where it arose almost a century ago in the work of É. Cartan on the classification of symmetric spaces. Here one considers a *symmetric Lie group* (G, τ), i.e., a Lie group G, endowed with an involutive automorphism τ. Then the Lie algebra $\mathfrak{g}$ of G decomposes into τ-eigenspaces

$$\mathfrak{h} = \{x \in \mathfrak{g} \colon \tau x = x\} \quad \text{and} \quad \mathfrak{q} = \{x \in \mathfrak{g} \colon \tau x = -x\}.$$

From the bracket relations $[\mathfrak{h}, \mathfrak{h}] \subseteq \mathfrak{h}$, $[\mathfrak{h}, \mathfrak{q}] \subseteq \mathfrak{q}$, and $[\mathfrak{q}, \mathfrak{q}] \subseteq \mathfrak{h}$ it then follows that the *Cartan dual*

$$\mathfrak{g}^c := \mathfrak{h} + i\mathfrak{q}$$

also is a Lie subalgebra of the complexified Lie algebra $\mathfrak{g}_{\mathbb{C}} = \mathfrak{g} \oplus i\mathfrak{g}$. We thus obtain a duality relation between symmetric Lie groups (G, τ) and (G^c, τ^c), where G^c denotes a Lie group with Lie algebra G^c and τ^c an involutive automorphism acting by $x + iy \mapsto x - iy$ on the Lie algebra $\mathfrak{g}^c = \mathfrak{h} + i\mathfrak{q}$. The classical examples from quantum physics are the euclidean motion group $G = E(d) \cong \mathbb{R}^d \rtimes O_d(\mathbb{R})$ and the automorphism τ of $E(d)$ induces by time reflection. This establishes a duality with the Poincaré group $G^c = P(d) \cong \mathbb{R}^{1,d-1} \rtimes O_{1,d-1}(\mathbb{R})$.

In many cases both groups G and G^c are contained in one complex Lie group $G_{\mathbb{C}}$ and $H = G \cap G^c$ is a Lie subgroup with Lie algebra $\mathfrak{h}$, contained in both. Therefore any passage from G to G^c should be related to analytic continuation to domains in $G_{\mathbb{C}}$ whose closure intersects both groups G and G^c. On the Lie algebra level the passage from $\mathfrak{g} = \mathfrak{h} \oplus \mathfrak{q}$ to $\mathfrak{g}^c = \mathfrak{h} \oplus i\mathfrak{q}$ very much resembles Wick rotation because the elements of $\mathfrak{q}$ are multiplied by i (cf. [HH17], where this context is discussed for pseudo-Riemannian manifolds). Simple examples are

K.-H. Neeb and G. Ólafsson, *Reflection Positivity*, SpringerBriefs in Mathematical Physics, https://doi.org/10.1007/978-3-319-94755-6_1

- the circle group $G = \mathbb{T} \subseteq G_{\mathbb{C}} = \mathbb{C}^\times$ with the dual group $G^c = \mathbb{R}^\times$ and $\tau(z) = z^{-1}$.
- the additive group $G = \mathbb{R} \subseteq G_{\mathbb{C}} = \mathbb{C}$ with $\tau(x) = -x$ and $G^c = i\mathbb{R}$.
- the group $G = \mathrm{GL}_n(\mathbb{R}) \subseteq G_{\mathbb{C}} = \mathrm{GL}_n(\mathbb{C})$ with $\tau(g) = g^{-\top}$ and $G^c = \mathrm{U}_n(\mathbb{C})$.
- the group $G = \mathrm{O}_n(\mathbb{R}) \subseteq G_{\mathbb{C}} = \mathrm{O}_n(\mathbb{C})$ with $\tau(g) = rgr$, where r is an orthogonal reflection in a hyperplane and $G^c = \mathrm{O}_{1,n-1}(\mathbb{R})$.
- the group $G = \mathrm{O}_{1,n}(\mathbb{R}) \subseteq G_{\mathbb{C}} = \mathrm{O}_{n+1}(\mathbb{C})$ with $\tau(g) = \theta g \theta$, where θ is an orthogonal reflection in a Minkowski hyperplane and $G^c = \mathrm{O}_{2,n-1}(\mathbb{R})$.

If $U: G \to \mathrm{U}(\mathcal{E})$ is a unitary representation of G, then, for $x \in \mathfrak{g}$, the infinitesimal generator $dU(x)$ of the unitary one-parameter $t \mapsto U(\exp tx)$ is a skew-adjoint operator, and multiplication by i leads to a selfadjoint operator. Therefore we cannot expect unitary representations of G and G^c to live on the same Hilbert space. What we need instead is some extra structure on $\mathcal{E}$ that permits us to construct another Hilbert space on which a unitary representation of G^c may be implemented. This is where reflection positivity comes in as a framework establishing a bridge between unitary representations of G and G^c. This perspective isolates many of the key features of reflection positivity and subsumes not only the representation theoretic aspects of classical applications along the lines of Osterwalder and Schrader [OS73, OS75], but also quite recent developments in Algebraic Quantum Field Theory (AQFT), where Haag–Kastler nets of operator algebras are constructed on space times by methods relying very much on the unitary or anti-unitary representations of the groups involved [Bo92, BJM16, NÓ17, Nel69]. Another recent branch of applications of reflection positivity for the euclidean conformal group along these lines concerns Hardy–Littlewood–Sobolev inequalities in analysis (see [FL10, FL11, NÓ14]).

The extra structure required on the Hilbert space $\mathcal{E}$ can be specified axiomatically as follows. A *reflection positive Hilbert space* is a triple $(\mathcal{E}, \mathcal{E}_+, \theta)$, consisting of a Hilbert space $\mathcal{E}$ with a unitary involution θ and a closed subspace $\mathcal{E}_+$ satisfying

$$\langle \xi, \xi \rangle_\theta := \langle \xi, \theta \xi \rangle \geq 0 \quad \text{for} \quad \xi \in \mathcal{E}_+.$$

This structure immediately leads to a new Hilbert space $\widehat{\mathcal{E}}$ that we obtain from the positive semidefinite form $\langle \cdot, \cdot \rangle_\theta$ on $\mathcal{E}_+$. We write $q: \mathcal{E}_+ \to \widehat{\mathcal{E}}, \xi \mapsto \widehat{\xi}$ for the natural map. Bounded or unbounded operators S on $\mathcal{E}_+$ preserving the kernel of q induce an operator $\widehat{S}$ on $\widehat{\mathcal{E}}$ via $\widehat{S}\widehat{\xi} := \widehat{S\xi}$. The passage $S \mapsto \widehat{S}$ is called the *Osterwalder–Schrader (OS) transform*.

On the level of $\mathcal{E}$ (the euclidean side), we consider a unitary representation U of a symmetric Lie group (G, τ) on a reflection positive Hilbert space $(\mathcal{E}, \mathcal{E}_+, \theta)$. There are several ways to express the compatibility of the representation U with $\mathcal{E}_+$ and θ. One is the compatibility relation

$$\theta U_g \theta = U_{\tau(g)} \quad \text{for} \quad g \in G$$

between τ and the unitary involution θ and another is the invariance of $\mathcal{E}_+$ under the operators U_h, where h belongs to the identity component $H := G_0^\tau$ of the group

G^τ of τ-fixed points in G. These two already ensure that the OS transform yields a unitary representation $(\widehat{U}_h)_{h \in H}$ on $\widehat{\mathscr{E}}$. We are aiming at a unitary representation of the Cartan dual group G^c on $\widehat{\mathscr{E}}$ and this can only be achieved by additional requirements. On the algebraic level, if we only consider the representation dU of the Lie algebra $\mathfrak{g}$ on the subspace $\mathscr{E}^\infty$ of smooth vectors for $(U, \mathscr{E})$, it suffices to have a subspace $\mathscr{D} \subseteq \mathscr{E}_+ \cap \mathscr{E}^\infty$ which is $\mathfrak{g}$-invariant. Then OS transformation immediately leads to a representation of $\mathfrak{g}^c = \mathfrak{h} + i\mathfrak{q}$ by skew-symmetric operators on the subspace $\widehat{\mathscr{D}} \subseteq \widehat{\mathscr{E}}$ by

$$x + iy \mapsto \big(dU(x)|_{\mathscr{D}} + i\, dU(y)|_{\mathscr{D}}\big)\widehat{\ }.$$

This simple passage already shows how the θ-twisting of the scalar product on $\mathscr{E}_+$ turns the symmetric operators $i\, dU(y)$, $y \in \mathfrak{q}$, on $\mathscr{D}$ into skew-symmetric operators on $\widehat{\mathscr{D}}$, but it completely ignores all issues related to essential selfadjointness and, accordingly, integrability to group representations. We therefore need more global ways to express the reflection positivity requirements.

One way to express such requirements refers to subsemigroups $S \subseteq G$ which are *symmetric* in the sense that they are invariant under the involution $s \mapsto s^\sharp := \tau(s)^{-1}$. Then we call $(U, \mathscr{E})$ *reflection positive with respect to* S if $U_s \mathscr{E}_+ \subseteq \mathscr{E}_+$ holds for $s \in S$. Then OS transformation yields a $*$-representation $(\widehat{U}_s)_{s \in S}$ of the involutive semigroup $(S, \sharp)$, and if S has interior points one can expect this representation to "extend analytically" to a unitary representation of a dual group G^c. A typical situation arises for $(G, S, \tau) = (\mathbb{R}, \mathbb{R}_+, -\,\mathrm{id}_\mathbb{R})$ in euclidean field theory from the one-parameter group of time translations. Then $(\widehat{U}_t)_{t \geq 0}$ is a one-parameter group of hermitian contractions on $\widehat{\mathscr{E}}$, hence of the form $\widehat{U}_t = e^{-tH}$ for a positive selfadjoint operator H, and $U_t^c := e^{itH}$ defines a unitary representation of the dual group $G^c = i\mathbb{R}$ with positive spectrum (in QFT H corresponds to the Hamiltonian, the energy observable, which should be positive).

There are, however, many situations where there are no natural symmetric subsemigroups $S \subseteq G$, or some which do not have interior points, such as the subsemigroup S of the euclidean motion group mapping a closed half space into itself. In this case the reflection positivity requirements on $(U, \mathscr{E}, \mathscr{E}_+, \theta)$ have to be formulated differently. Instead of a subsemigroup, we consider a domain $G_+ \subseteq G$ (mostly open or with dense interior) and the reflection positivity condition is inspired by situations in QFT, where Hilbert spaces are generated by field operators: Instead of fixing $\mathscr{E}_+$ a priori, we consider a real linear space V and a linear map $j : V \to \mathscr{H}$ whose range generates $\mathscr{E}$ under U_G and θ and call $(U, \mathscr{E}, j, V)$ *reflection positive with respect to* G_+ if the subspace $\mathscr{E}_+ := [\![U_{G_+}^{-1} j(V)]\!]$ is θ-positive. The prototypical examples arise for circle groups $G = \mathbb{R}/\beta\mathbb{Z}$ with $\tau(g) = g^{-1}$, where $G_+ = [0, \pi] + \beta\mathbb{Z}$ is a half circle. In physics they occur in the context of quantum statistical mechanics, where β plays the role of an inverse temperature [Fro11]. Both approaches, the one based on semigroups S and on domains G_+ lead to situations in which we can use suitable integrability results (cf. Chap. 6) to obtain unitary representations of the 1-connected Lie group G^c with Lie algebra $\mathfrak{g}^c = \mathfrak{h} + i\mathfrak{q}$ on $\widehat{\mathscr{E}}$.

We now turn to the contents of this book. We shall not turn to finer aspects of unitary representations of higher-dimensional Lie groups before Chap. 6. In the first half, Chaps. 2–5, we deal with rather concrete contexts and how they relate to reflection positivity. Various aspects concerning general Lie groups are postponed to Chaps. 6–9.

Chapter 2 develops the notion of a reflection positive Hilbert space $(\mathscr{E}, \mathscr{E}_+, \theta)$ from various perspectives. For instance θ-positive subspaces $\mathscr{E}_+$ can be constructed as graphs of contractions from the 1 to the -1-eigenspace of θ (Sect. 2.2). In physics reflection positive Hilbert spaces often arise from distributions. Here $\mathscr{E}$ is a Hilbert space arising by completing the space $C_c^\infty(M)$ of smooth test functions on a manifold with respect to a singular scalar product

$$\langle \xi, \eta \rangle = \int_{M \times M} \overline{\xi(x)} \eta(y) \, dD(x, y), \tag{1.1}$$

where D is a positive definite distribution on $M \times M$. Then θ is supposed to come from a diffeomorphism of M and $\mathscr{E}_+$ from an open subset $M_+ \subseteq M$, which leads to the reflection positivity condition

$$\int_{M_+ \times M_+} \overline{\xi(x)} \xi(y) \, dD(\theta(x), y) \geq 0 \quad \text{for} \quad \xi \in C_c^\infty(M_+) \tag{1.2}$$

(Section 2.4). Typical concrete examples arise from reflections of complete Riemannian manifolds and resolvents $(\lambda \mathbf{1} - \Delta)^{-1}$ of the Laplacian (Sect. 2.5). Motivated by these examples, we briefly describe an abstract operator theoretic context for reflection positivity that we feel should be developed further (Sect. 2.6; [JR07, An13, AFG86, Di04]). In probabilistic contexts, one encounters situations satisfying the Markov condition, i.e., there exists a subspace $\mathscr{E}_0 \subseteq \mathscr{E}_+$ mapped isometrically onto $\widehat{\mathscr{E}}$ (Sect. 2.3).

The connection between reflection positive Hilbert spaces and representation theory is introduced in Chap. 4. After discussing some general properties of the OS transform, we introduce symmetric Lie groups (G, τ), symmetric subsemigroups $S \subseteq G$ and various kinds of reflection positivity conditions for unitary representations. As in the representation theory of operator algebras, where cyclic representations are generated from states, it is an extremely fruitful approach to generate representations of groups and semigroups by positive definite functions via the Gelfand–Naimark–Segal (GNS) construction. Here reflection positivity requirements lead to the concept of a reflection positive function whose values may also comprise bounded operators or bilinear forms (Sect. 3.4).

After these generalities, we turn in Chap. 4 to the most elementary concrete symmetric Lie group $(G, \tau) = (\mathbb{R}, -\mathrm{id}_\mathbb{R})$, where the RP condition is based on the subsemigroup $\mathbb{R}_+$. Although this Lie group is quite trivial, reflection positivity on the real line has many interesting facets and is therefore quite rich. As reflection positive functions play a crucial role, we start Chap. 4 with functions on intervals $(-a, a) \subseteq \mathbb{R}$ which are reflection positive in the sense that both kernels

$$(\varphi(x - y))_{-a/2 < x, y < a/2} \quad \text{and} \quad (\varphi(x + y))_{0 < x, y < a/2}$$

are positive definite. For $a = \infty$, this combines the positive definiteness of the group $(\mathbb{R}, +)$ with the positive definiteness on the involutive semigroup $(\mathbb{R}_+, \mathrm{id}_\mathbb{R})$. Accordingly, these two conditions ask for techniques related to Fourier- and Laplace transforms.

Reflection positive representations of $(\mathbb{R}, \mathbb{R}_+, -\mathrm{id}_\mathbb{R})$ are unitary one-parameter groups $(U_t)_{t \in \mathbb{R}}$ on a reflection positive Hilbert space $(\mathscr{E}, \mathscr{E}_+, \theta)$ satisfying $U_t \mathscr{E}_+ \subseteq \mathscr{E}_+$ for $t > 0$ and $\theta U_t \theta = U_{-t}$ for $t \in \mathbb{R}$. On $\widehat{\mathscr{E}}$ this leads to a semigroup $(\widehat{U}_t)_{t \geq 0}$ of hermitian contractions and we show in particular that, under the OS transform, fixed points of U on $\mathscr{E}$ correspond to fixed points of $\widehat{U}$ in $\widehat{\mathscr{E}}$ (Proposition 4.2.6).

For $(\mathbb{R}, \mathbb{R}_+, -\mathrm{id}_\mathbb{R})$, we obtain a complete classification of reflection positive representations in terms of integral formulas, resp., spectral theorems. From these results one obtains an interesting converse of the OS transform in this context. Any hermitian contraction semigroup $(C_t)_{t \geq 0}$ on a Hilbert space $\mathscr{H}$ has a so-called minimal dilation represented by the reflection positive function $\psi(t) := C_{|t|}$ on $\mathbb{R}$.

We conclude Chap. 4 by showing that, for any reflection positive one-parameter group for which $\mathscr{E}_+$ is cyclic and fixed points are trivial, the space $\mathscr{E}_+$ is outgoing in the sense of Lax–Phillips scattering theory (Proposition 4.4.2). This establishes a remarkable connection between reflection positivity and scattering theory that leads to a normal form of reflection positive one-parameter groups by translations on spaces of the form $\mathscr{E} = L^2(\mathbb{R}, \mathscr{H})$ with $\mathscr{E}_+ = L^2(\mathbb{R}_+, \mathscr{H})$. Applying the Fourier transform to our concrete dilation model leads precisely to this normal form.

In Chap. 5 we still work with the same symmetric group $(\mathbb{R}, -\mathrm{id}_\mathbb{R})$ or rather its quotient circle group $\mathbb{R}/2\beta\mathbb{Z} \cong \mathbb{T}$, but now reflection positivity is based on the interval $[0, \beta]$, where $\beta > 0$ is interpreted as an inverse temperature in physical models [NÓ15b, KL81, KL81b, Fro11, NÓ16]. In this context reflection positivity is closely connected with the Kubo–Martin–Schwinger (KMS) condition for states of C^*-dynamical systems [Fro11, BR02]. This connection is established by a purely representation theoretic perspective on the KMS condition formulated as a property of form-valued positive definite functions on $\mathbb{R}$: Let V be a real vector space and $\mathrm{Bil}(V)$ be the space of real bilinear maps $V \times V \to \mathbb{C}$. For $\beta > 0$, we consider the open strip

$$\mathscr{S}_\beta := \{z \in \mathbb{C} \colon 0 < \mathrm{Im}\, z < \beta\}.$$

We say that a positive definite function $\psi \colon \mathbb{R} \to \mathrm{Bil}(V)$ (Definition A.1.5) satisfies the β-KMS condition if ψ extends to a pointwise continuous function ψ on $\overline{\mathscr{S}_\beta}$ which is pointwise holomorphic on the interior $\mathscr{S}_\beta$ and satisfies

$$\psi(i\beta + t) = \overline{\psi(t)} \quad \text{for} \quad t \in \mathbb{R}.$$

The classification of such functions in terms of an integral representation is based on relating them to standard (real) subspaces of a complex Hilbert space which occur naturally in the modular theory of operator algebras [Lo08]. These are closed real subspaces $V \subseteq \mathcal{H}$ for which $V \cap iV = \{0\}$ and $V + iV$ is dense. Any standard subspace determines a pair (Δ, J) of *modular objects*, where Δ is a positive selfadjoint operator and J is an anti-linear involution (a *conjugation*) satisfying $J\Delta J = \Delta^{-1}$. The connection is established by

$$V = \text{Fix}(J\Delta^{1/2}) = \{\xi \in \mathcal{D}(\Delta^{1/2}) : J\Delta^{1/2}\xi = \xi\}.$$

This connects reflection positivity very naturally to the aforementioned recent developments in AQFT initiated by the work of Borchers [Bo92] and now exploited systematically in the constructions of QFTs (see [BJM16, BLS11, LL14, LW11] for typical applications). Here standard subspaces can be considered as "one-particle space analogs" of the modular data (Δ, J) arising in Tomita–Takesaki theory in the context of von Neumann algebras [Lo08, LL14].

After the discussion of the concrete examples, the reader should be prepared to appreciate the Lie theoretic aspects of the theory, which start in Chap. 6 with the development of the integration techniques that are used to obtain a unitary representation of the simply connected Lie group G^c on $\widehat{\mathscr{E}}$ from a reflection positive representation of (G, τ) on $(\mathscr{E}, \mathscr{E}_+, \theta)$. Our techniques are based on the fact that the Hilbert spaces are mostly constructed from G-invariant positive definite kernels or positive definite G-invariant distributions. We have already seen that any reflection positive representation of (G, τ) immediately yields a unitary representation U^c of $H = G_0^\tau$ on $\widehat{\mathscr{E}}$, so that it remains to find a unitary representation of the one-parameter groups $\exp_{G^c}(\mathbb{R}ix)$ for $x \in \mathfrak{q}$. By Stone's Theorem, the main point is to show that, for $y \in \mathfrak{q}$, the symmetric operator $\widehat{\mathrm{d}U}(y)$ defined on a dense subspace of $\widehat{\mathscr{E}}$ is essentially selfadjoint. In our geometric setting, this can be derived from Fröhlich's Theorem [Fro80] which provides a criterion for the essential selfadjointness of a symmetric operator in terms of the existence of enough local solutions of the corresponding linear ODE. The natural setting for the corresponding integrability results are pairs (β, σ) of a homomorphism $\beta \colon \mathfrak{g} \to \mathcal{V}(M)$ to the Lie algebra $\mathcal{V}(M)$ of smooth vector fields on a manifold M which is compatible with a smooth H-action σ. Then, for any smooth kernel K on M satisfying a suitable invariance condition with respect to (β, σ), a unitary representation of G^c on $\mathscr{H}_K$ exists (Theorem 6.2.3). We also show that a similar result holds if we replace the kernel K by a positive definite distribution $K \in C^{-\infty}(M \times M)$ compatible with (β, σ) (Theorem 6.3.6). From this we easily derive the existence of a unitary representation of the simply connected group G^c on $\widehat{\mathscr{E}}$ for a reflection positive representation $(U, \mathscr{E})$ of (G, τ). Our exposition is based on new aspects developed in [MNO15] which complements the classical approach from [FOS83].

The most effective tool to deal with reflection positive representations of symmetric Lie groups (G, τ) are reflection positive distributions on G and their relation with reflection positive distribution vectors of unitary representations. A key advantage of this method, outlined in Chap. 7, is that is leads naturally to reflection positive representations in Hilbert spaces of distributions on homogeneous spaces G/H, where H may be non-compact. To illustrate this technique, we apply it to spherical representations of the Lorentz group $G = O_{1,n}(\mathbb{R})$. These representations consist of two series, the principal series and the complementary series. Both have natural realizations in spaces of distributions on the n-sphere $\mathbb{S}^n \cong G/P$ on which the Lorentz group acts by conformal maps; the principal series can even be realized in $L^2(\mathbb{S}^n)$.

In Chap. 8 we take a closer look at the representations of the Poincaré group corresponding to scalar generalized free fields and their euclidean realizations by representations of the euclidean motion group $E(d)$. In particular, we discuss Lorentz invariant measures on the forward light cone $\overline{V_+}$ and the corresponding unitary representations of the Poincaré group. Applying the dilation construction to the time translation semigroup leads immediately to a Hilbert space $\mathscr{E}$ on which we have a unitary representation of the euclidean motion group. We also characterize those representations which extend to the conformal group $O_{2,d}(\mathbb{R})$ of Minkowski space $\mathbb{R}^{1,d-1}$. Then the euclidean realization is a unitary representation of the Lorentz group $O_{1,d+1}(\mathbb{R})$, acting as the conformal group on euclidean $\mathbb{R}^d$.

A particularly fascinating aspect of reflection positivity is its intimate connection with stochastic processes which is briefly scratched in Chap. 9. This is already interesting in the context of one-parameter groups, where it surfaces for example in the fact that the unitary one-parameter group $(U_t)_{t \in \mathbb{R}}$ leads by OS transform to the one-parameter group $U_t^c = e^{-it\Delta}$ on $L^2(\mathbb{R}^n)$, respectively to the heat semigroup $e^{t\Delta}$, is the translation action of $\mathbb{R}$ on a suitable Lebesgue–Wiener space. This connection was observed by Nelson in [Nel64] and led to a new approach to Feynman–Kac type integral formulas. In Chap. 9 we describe some recent generalizations of classical results of Klein and Landau [Kl78, KL75] concerning the interplay between reflection positivity and stochastic processes. Here the main step is the passage from the symmetric semigroup $(\mathbb{R}, \mathbb{R}_+, -\mathrm{id}_{\mathbb{R}})$ to a more general context (G, S, τ). This leads to the concept of a (G, S, τ)-measure space generalizing Klein's Osterwalder–Schrader path spaces for $(\mathbb{R}, \mathbb{R}_+, -\mathrm{id}_{\mathbb{R}})$. A key result of this theory is the correspondence between (G, S, τ)-measure spaces and the corresponding positive semigroup structures on the Hilbert space $\widehat{\mathscr{E}}$.

Notation

We write $\mathbb{R}_{\geq 0} := [0, \infty)$ for the closed half line, $\mathbb{R}_+ = (0, \infty)$ for the open half line and $\mathbb{N} = \{1, 2, 3, \ldots\}$ for the set of natural numbers.

As customary in physics, all scalar products on Hilbert spaces $\mathscr{H}$ will be linear in the second argument. A subset S of $\mathscr{H}$ is called *total* if it spans a dense subspace. We write $[\![S]\!] := \overline{\operatorname{span} S}$.

We write $U(\mathscr{H})$ for the unitary group of a Hilbert space $\mathscr{H}$.

For a measure space $(X, \mathfrak{S}, \mu)$, we accordingly write

$$\langle f, g \rangle = \int_X \overline{f(x)} g(x) \, d\mu(x) \quad \text{for} \quad f, g \in L^2(X, \mu).$$

We write elements of $\mathbb{R}^d$ as $x = (x_0, x_1, \ldots, x_{d-1}) = (x_0, \mathbf{x})$. The standard inner product on $\mathbb{R}^d$ is denoted $\langle x, y \rangle = x \cdot y = xy = \sum_{j=0}^{d-1} x_j y_j$, and the Lorentzian inner product by

$$[x, y] = x_0 y_0 - \mathbf{x}\mathbf{y}.$$

In d-dimensional Minkowski space M^d we write

$$V_+ := \{p = (p_0, \mathbf{p}) \in \mathbb{R}^d : p_0 > 0, \, p_0^2 > \mathbf{p}^2\}$$

for the open *forward light cone*.

We write $C_c^\infty(M)$ for the space of complex-valued test functions and $\mathscr{S}(\mathbb{R}^d)$ is the space of complex-valued Schwartz functions on $\mathbb{R}^d$. For the *Fourier transform of a measure* μ on the dual V^* of a finite-dimensional real vector space V, we write

$$\widehat{\mu}(x) := \int_{V^*} e^{-i\alpha(x)} \, d\mu(\alpha).$$

The *Fourier transform of an L^1-function* f on $\mathbb{R}^d$ is defined by

$$\widehat{f}(p) := \int_{\mathbb{R}^d} e^{-i\langle p, x\rangle} f(x) \, d\lambda_d(x) = \frac{1}{(2\pi)^{d/2}} \int_{\mathbb{R}^d} e^{-i\langle p, x\rangle} f(x) \, dx \qquad (1.3)$$

which corresponds to the Fourier transform of the measure $f\lambda_d$, where $d\lambda_d(x) := (2\pi)^{-d/2} \cdot dx$ is a suitably normalized Lebesgue measure. We likewise define convolution of L^1-functions with respect to λ_d.

For tempered distributions $D \in \mathscr{S}'(\mathbb{R}^d)$, which we define as continuous antilinear functionals on the Schwartz space $\mathscr{S}(\mathbb{R}^d)$, we define the Fourier transform by

$$\widehat{D}(\varphi) := D(\widetilde{\varphi}), \quad \text{where} \quad \widetilde{\varphi}(p) = \widehat{\varphi}(-p) = \int_{\mathbb{R}^d} e^{i\langle p, x\rangle} \varphi(x) \, d\lambda_d(x). \qquad (1.4)$$

For $D_f(\varphi) = \int_{\mathbb{R}^d} \overline{\varphi(x)} f(x) \, d\lambda_d(x)$, we then have $\widehat{D_f} = D_{\widehat{f}}$ and for a tempered measure μ the corresponding distribution $D_\mu(\varphi) := \int \overline{\varphi} \, d\mu$ satisfies $\widehat{D_\mu} = D_{\widehat{\mu}}$ if we consider $\widehat{\mu}$ as a function. For the point measure δ_0 we then have in particular the relation

$$1 = \widehat{\delta_0}$$

which corresponds to the normalized Lebesgue measure λ_d.

Chapter 2
Reflection Positive Hilbert Spaces

In this chapter we discuss the basic framework of reflection positivity: reflection positive Hilbert spaces. These are triples $(\mathscr{E}, \mathscr{E}_+, \theta)$, consisting of a Hilbert space $\mathscr{E}$, a unitary involution θ on $\mathscr{E}$ and a closed subspace $\mathscr{E}_+$ which is θ-*positive* in the sense that $\langle \xi, \theta \xi \rangle \geq 0$ for $\xi \in \mathscr{E}_+$. This structure immediately leads to a new Hilbert space $\widehat{\mathscr{E}}$ and a linear map $q \colon \mathscr{E}_+ \to \widehat{\mathscr{E}}$ with dense range. When the so-called Markov condition is satisfied, there even exists a closed subspace $\mathscr{E}_0 \subseteq \mathscr{E}_+$ mapped isometrically onto $\widehat{\mathscr{E}}$ (Sect. 2.3). Reflection positive Hilbert spaces arise naturally in many different contexts: as graphs of contractions (Sect. 2.2), from reflection positive distribution kernels on manifolds (Sect. 2.4) and in particular from dissecting reflections of complete Riemannian manifolds and resolvents of the Laplacian (Sect. 2.5). This motivates the short discussion of an abstract operator theoretic context of reflection positivity in Sect. 2.6.

2.1 Reflection Positive Hilbert Spaces

We start with the definition of a reflection positive Hilbert space:

Definition 2.1.1 (*Reflection positive Hilbert space*) Let $\mathscr{E}$ be a real or complex Hilbert space and $\theta \in \mathrm{U}(\mathscr{E})$ be a unitary involution. A closed subspace $\mathscr{E}_+ \subseteq \mathscr{E}$ is called θ-*positive* if $\langle \eta, \theta \eta \rangle \geq 0$ for $\eta \in \mathscr{E}_+$. We then call the triple $(\mathscr{E}, \mathscr{E}_+, \theta)$ a *reflection positive Hilbert space*.

If $(\mathscr{E}, \mathscr{E}_+, \theta)$ is a reflection positive Hilbert space, then

$$\mathscr{N} := \{\eta \in \mathscr{E}_+ \colon \langle \eta, \theta \eta \rangle = 0\} = \{\eta \in \mathscr{E}_+ \colon (\forall \zeta \in \mathscr{E}_+) \, \langle \zeta, \theta \eta \rangle = 0\}$$

is the subspace of $\mathscr{E}_+$ on which the new scalar product

© The Author(s) 2018
K.-H. Neeb and G. Ólafsson, *Reflection Positivity*, SpringerBriefs in Mathematical Physics, https://doi.org/10.1007/978-3-319-94755-6_2

$$\langle \eta, \xi \rangle_\theta := \langle \eta, \theta \xi \rangle$$

degenerates. We thus consider the quotient map

$$q : \mathcal{E}_+ \to \mathcal{E}_+ / \mathcal{N}, \quad \eta \mapsto \widehat{\eta} \tag{2.1}$$

and write $\widehat{\mathcal{E}}$ for the Hilbert space completion of $\mathcal{E}_+ / \mathcal{N}$ with respect to the norm $\|\widehat{\eta}\|_{\widehat{\mathcal{E}}} := \|\widehat{\eta}\| := \sqrt{\langle \eta, \theta \eta \rangle}$.

2.2 Reflection Positive Subspaces as Graphs

To get a better picture of how reflection positive Hilbert spaces arise, we now describe a construction of these structures in terms of contractions on a subspace of $\mathcal{E}^\theta$.

We start with a unitary involution θ on the Hilbert space $\mathcal{E}$. Then θ is diagonalizable with the two eigenspaces $\mathcal{E}_{\pm 1} := \{\eta \in \mathcal{E} : \theta \eta = \pm \eta\}$ and $\mathcal{E} = \mathcal{E}_{+1} \oplus \mathcal{E}_{-1}$. Then the twisted inner product $\langle \cdot, \cdot \rangle_\theta$ is positive definite on $\mathcal{E}_{+1}$ and negative definite on $\mathcal{E}_{-1}$:

$$\langle \xi_+ + \xi_-, \xi_+ + \xi_- \rangle_\theta = \|\xi_+\|^2 - \|\xi_-\|^2 \quad \text{for} \quad \xi_\pm \in \mathcal{E}_{\pm 1}. \tag{2.2}$$

Denote by $p_\pm$ the projection onto $\mathcal{E}_{\pm 1}$. Let $\mathcal{E}_+ \subseteq \mathcal{E}$ be a θ-positive subspace and $\mathcal{F} := p_+(\mathcal{E}_+)$ be its projection onto $\mathcal{E}_{+1}$. Then $\mathcal{E}_+ \cap \mathcal{E}_{-1} = \{0\}$ implies that there exists a linear map $C : \mathcal{F} \to \mathcal{E}_{-1}$ such that

$$\mathcal{E}_+ = \mathcal{G}(C) = \{u + Cu : u \in \mathcal{F}\}$$

is the graph of C. Now (2.2) yields $\langle u + Cu, u + Cu \rangle_\theta = \|u\|^2 - \|Cu\|^2 \geq 0$ for $u \in \mathcal{F}$, so that $\|C\| \leq 1$, i.e., C is a contraction. If, conversely, $\mathcal{F} \subseteq \mathcal{E}_{+1}$ is a subspace and $C : \mathcal{F} \to \mathcal{E}_{-1}$ is a contraction, then its graph $\mathcal{G}(C) \subseteq \mathcal{E}_{+1} \oplus \mathcal{E}_{-1} = \mathcal{E}$ is θ-positive. Since $\mathcal{G}(C)$ is closed if and only if $\mathcal{F}$ is closed, we obtain the following lemma which provides a description of all θ-positive subspaces in terms of contractions (cf. [JN16, Lemma 5.1]):

Lemma 2.2.1 *A closed subspace $\mathcal{E}_+ \subseteq \mathcal{E}$ is θ-positive if and only if there exists a closed subspace $\mathcal{F} \subseteq \mathcal{E}_{+1}$ and a contraction $C : \mathcal{F} \to \mathcal{E}_{-1}$ such that $\mathcal{E}_+ = \mathcal{G}(C)$.*

Remark 2.2.2 Let $(\mathcal{E}, \mathcal{E}_+, \theta)$ be a reflection positive Hilbert space.

(a) Put $\mathcal{E}_- := \theta(\mathcal{E}_+)$. Then $\mathcal{E}_+ \cap \mathcal{E}_-$ is the maximal θ-invariant subspace of $\mathcal{E}_+$, and θ-positivity of $\mathcal{E}_+$ implies that it coincides with $\mathcal{E}_0 := \{v \in \mathcal{E}_+ : \theta v = v\}$. This is the maximal subspace of $\mathcal{E}_+$ on which q is isometric.

(b) For $\mathcal{E}_+ = \mathcal{G}(C)$ as in Lemma 2.2.1, we have $\mathcal{E}_0 = \mathcal{E}_+ \cap \mathcal{E}_{+1} = \ker(C)$. In particular, $\mathcal{E}_0 = \{0\}$ if and only if C is injective.

(c) Writing $\mathcal{E}$ as $\mathcal{E}_0 \oplus \mathcal{E}_1$ with $\mathcal{E}_1 := \mathcal{E}_0^\perp$, the reflection positive Hilbert space is a direct sum of the trivial reflection positive Hilbert space $(\mathcal{E}_0, \mathcal{E}_0, \mathrm{id})$ and the reflection positive Hilbert space $(\mathscr{E}_1, \mathscr{E}_{1,+}, \theta_1)$, where $\theta_1 := \theta|_{\mathscr{E}_1}$, $\mathscr{E}_{1,+} = \mathscr{E}_1 \cap \mathscr{E}_+$, and $\mathscr{E}_{1,0} = \{0\}$.

(d) If $(\mathscr{E}, \mathscr{E}_+, \theta)$ is a reflection positive Hilbert space, then $(\mathscr{E}, \mathscr{E}_-, \theta)$ is reflection positive as well. If $\mathscr{E}_+ = \mathscr{G}(C)$ as above, then $\mathscr{E}_- = \mathscr{G}(-C)$.

2.3 The Markov Condition

For a reflection positive Hilbert space $(\mathscr{E}, \mathscr{E}_+, \theta)$, (2.2) shows that the subspace

$$\mathscr{E}_0 = \{\xi \in \mathscr{E}_+ : \theta\xi = \xi\}$$

is maximal with respect to the property that $q|_{\mathscr{E}_0} : \mathscr{E}_0 \to \widehat{\mathscr{E}}$ is isometric. In particular, $\mathscr{E}_0 \cong q(\mathscr{E}_0)$ is a closed subspace of $\widehat{\mathscr{E}}$.

An interesting special case arises if $\widehat{\mathscr{E}} = q(\mathscr{E}_0)$. Then q restricts to a unitary operator $\mathscr{E}_0 \to \widehat{\mathscr{E}}$, so that $q : \mathscr{E}_+ \to \widehat{\mathscr{E}}$ is a partial isometry with kernel $\mathscr{N} = \mathscr{E}_+ \ominus \mathscr{E}_0$. The following lemma characterizes this situation in terms of the *Markov condition* that originally arose in the context of stochastic processes (cf. Chap. 9).

Definition 2.3.1 Let $(\mathscr{E}, \mathscr{E}_+, \theta)$ be a reflection positive Hilbert space. If $\mathscr{E}_0' \subseteq \mathscr{E}_0$ is a closed subspace, $\mathscr{E}_- := \theta(\mathscr{E}_+)$, and P_0, $P_\pm$ are the orthogonal projections onto $\mathscr{E}_0'$ and $\mathscr{E}_\pm$, then we say that $(\mathscr{E}, \mathscr{E}_0', \mathscr{E}_+, \theta)$ is a *reflection positive Hilbert space of Markov type* if

$$P_+ P_0 P_- = P_+ P_-. \tag{2.3}$$

Lemma 2.3.2 *The Markov condition* (2.3) *is equivalent to* $\mathscr{E}_0' = \mathscr{E}_0$ *and* $q(\mathscr{E}_0) = \widehat{\mathscr{E}}$. *If it is satisfied, then*

(a) $\Gamma := q|_{\mathscr{E}_0} : \mathscr{E}_0 \to \widehat{\mathscr{E}}$ *is a unitary isomorphism and* $q = \Gamma \circ P_0|_{\mathscr{E}_+}$.
(b) *If* $\mathscr{E}_+ + \mathscr{E}_-$ *is dense in* $\mathscr{E}$, *then* $\mathscr{E}_+$ *is maximal* θ-*positive.*

Proof If $\mathscr{E}_0' = \mathscr{E}_0$ and $q(\mathscr{E}_0) = \widehat{\mathscr{E}}$, then $\mathscr{N} = \mathscr{E}_+ \cap \mathscr{E}_-^\perp = \ker q$ implies that $\mathscr{N} = \mathscr{E}_+ \ominus \mathscr{E}_0$. This leads to the orthogonal decomposition $\mathscr{E}_+ + \mathscr{E}_- = \theta(\mathscr{N}) \oplus \mathscr{E}_0 \oplus \mathscr{N}$ and to $\mathscr{E}_0 = \mathscr{E}_+ \cap \mathscr{E}_-$. Therefore $P_+ P_0 P_- = P_0 = P_+ P_-$.

Suppose, conversely, that the Markov condition holds. For $u \in \mathscr{E}_0 \subseteq \mathscr{E}_-$, we have $P_+ P_- u = u$, but $P_+ P_0 P_- u = P_0 u$, so that $\mathscr{E}_0 = \mathscr{E}_0'$. As $\mathscr{E}_0 \subset \ker(\theta - \mathbf{1})$, we have $P_0 \theta = \theta P_0 = P_0$. This implies

$$P_+ \theta P_+ = P_+ P_- \theta = P_+ P_0 P_- \theta = P_0 \theta P_+ = P_0 P_+ = P_0.$$

For $u \in \mathscr{E}_+$, we thus obtain $\langle u, \theta u \rangle = \langle u, P_+ \theta P_+ u \rangle = \langle u, P_0 u \rangle = \|P_0 u\|^2$. Therefore $\mathscr{N} = \ker q = \mathscr{E}_+ \ominus \mathscr{E}_0$ and $q(\mathscr{E}_0) = \widehat{\mathscr{E}}$. The remaining assertions are now clear. This implies the first assertion and (a).

We now verify (b). From $\mathcal{N} = \mathcal{E}_+ \ominus \mathcal{E}_0$ we derive that $\mathcal{E}_- = \mathcal{E}_0 \oplus \theta(\mathcal{N})$, so that $\mathcal{E}_+ + \mathcal{E}_- = \mathcal{E}_+ \oplus \theta(\mathcal{N})$ is an orthogonal decomposition and our assumption implies that $\mathcal{E} = \mathcal{E}_+ \oplus \theta(\mathcal{N})$. It follows that any proper enlargement $\widehat{\mathcal{E}}_+ \supsetneq \mathcal{E}_+$ contains a non-zero $\theta(\xi), \xi \in \mathcal{N}$. Then $\mathbb{C}\xi + \mathbb{C}\theta(\xi) \subseteq \widehat{\mathcal{E}}_+$ is θ-positive and θ-invariant, hence contained in $\mathcal{E}^\theta$, which contradicts the orthogonality of $\mathcal{N}$ and $\theta(\mathcal{N})$. □

Remark 2.3.3 (*Relation to stochastic processes*) Let $(X_t)_{t \in \mathbb{R}}$ be a full stochastic process on the probability space (Q, Σ, μ) (see Definition 9.2.1) and $\Sigma_t \subseteq \Sigma$ the smallest σ-subalgebra for which X_t is measurable. Accordingly, we define $\Sigma_\pm$ as the σ-subalgebra generated by all Σ_t for $\pm t \geq 0$. In $\mathcal{E} := L^2(Q, \Sigma, \mu)$ we thus obtain closed subspaces $\mathcal{E}_\pm := L^2(Q, \Sigma_\pm, \mu)$ and $\mathcal{E}_0 := L^2(Q, \Sigma_0, \mu)$. If $P_\pm$ and P_0 are the corresponding projections (corresponding to conditional expectations in this context), then the Markov condition (2.3) holds for all translates of the process $(X_t)_{t \in \mathbb{R}}$ if and only if it is a Markov process (cf. [JT17, Sect. 7]).

2.4 Reflection Positive Kernels and Distributions

There are many ways to specify Hilbert spaces concretely. Often they arise as L^2-spaces of measures, but here we shall mostly deal with spaces on which the inner product is specified differently, namely by a positive definite kernel. For detailed definitions and basic properties of positive definite kernels in various contexts, we refer to Appendix A.1.

Definition 2.4.1 Suppose that $K: X \times X \to \mathbb{C}$ is a positive definite kernel on the set X and that $\tau: X \to X$ is an involution leaving K invariant: $K(\tau x, \tau y) = K(x, y)$ for $x, y \in X$. If $X_+ \subseteq X$ is a subset with the property that the kernel

$$K^\tau: X_+ \times X_+ \to \mathbb{C}, \quad K^\tau(x, y) := K(x, \tau y) \tag{2.4}$$

is also positive definite, then we say that K is *reflection positive* with respect to (X, X_+, τ).

Lemma 2.4.2 *Let $K: X \times X \to \mathbb{C}$ be a kernel which is reflection positive with respect to (X, X_+, τ) and let $\mathcal{E} := \mathcal{H}_K \subseteq \mathbb{C}^X$ denote the corresponding reproducing kernel Hilbert space. Then the following assertions hold:*

(a) *$\theta f := f \circ \tau$ defines a unitary involution on $\mathcal{E}$.*
(b) *$\mathcal{E}_+ := [\![K_x : x \in X_+]\!]$ is a θ-positive subspace, so that $(\mathcal{E}, \mathcal{E}_+, \theta)$ is reflection positive.*
(c) *The map $\mathcal{E}_+ \to \mathbb{C}^{X_+}$, $f \mapsto f \circ \tau|_{X_+}$ induces a unitary isomorphism $\widehat{\mathcal{E}} \to \mathcal{H}_{K^\tau}$, so that we may identify $\widehat{\mathcal{E}}$ with the reproducing kernel space $\mathcal{H}_{K^\tau}$ and write $q(f) = f \circ \tau|_{X_+}$.*

Proof (a) The invariance of K under τ implies the existence of a unitary involution θ on $\mathcal{H}_K$ with $\theta(K_x) = K_{\tau x}$. Then $(\theta f)(x) = \langle K_x, \theta f \rangle = \langle K_{\tau x}, f \rangle = f(\tau x)$ shows that $\theta f = f \circ \tau$.

(b) For $x, y \in X_+$, we have $\langle K_x, \theta K_y \rangle = \langle K_x, K_{\tau y} \rangle = K(x, \tau y) = K^\tau(x, y)$, and this implies that the closed subspace $\mathscr{E}_+ \subseteq \mathscr{E}$ generated by $(K_x)_{x \in X_+}$ is θ-positive.

(c) The Hilbert space $\widehat{\mathscr{E}}$ is generated by the elements $q(K_x)$, $x \in X_+$, and we have

$$\langle q(K_x), q(K_y) \rangle_{\widehat{\mathscr{E}}} = \langle K_x, \theta K_y \rangle = K^\tau(x, y).$$

This implies that $\widehat{\mathscr{E}} \cong \mathscr{H}_{K^\tau}$ and that the function on X_+ corresponding to $q(f) \in \widehat{\mathscr{E}}$ is given by $q(f)(x) = \langle q(K_x), q(f) \rangle_{\widehat{\mathscr{E}}} = \langle K_x, \theta f \rangle = f(\tau x)$. $\qquad\square$

All reflection positive spaces can be construction from reflection positive kernels: If $(\mathscr{E}, \mathscr{E}_+, \theta)$ is a reflection positive Hilbert space, then the scalar product defines a reflection positive kernel $K(\eta, \zeta) := \langle \eta, \zeta \rangle$ on $X = \mathscr{E}$, and this kernel is reflection positive with respect to $(\mathscr{E}, \mathscr{E}_+, \theta)$.

Example 2.4.3 On $X = \mathbb{R}$, we consider the involution $\tau(x) = -x$.

(a) We claim that, for every $\lambda \geq 0$, the kernel $K(x, y) = e^{-\lambda|x-y|}$ is reflection positive with respect to $(\mathbb{R}, \mathbb{R}_+, \theta)$.
   The positive definiteness of K means that the function $\varphi_\lambda(x) := e^{-\lambda|x|}$ (multiples of euclidean Green's functions [DG13]) is a positive definite function on the group $(\mathbb{R}, +)$. In view of Bochner's Theorem (Theorem A.2.1), this is equivalent to φ_λ being the Fourier transform of a positive measure. In fact,

$$\varphi_\lambda(x) = e^{-\lambda|x|} = \int_{\mathbb{R}} e^{-ixp} \, d\mu_\lambda(p), \quad \text{where} \quad d\mu_\lambda(p) = \frac{\lambda}{\pi} \frac{dp}{\lambda^2 + p^2} \quad (2.5)$$

is the Cauchy distribution. To verify reflection positivity, we observe that, for $x, y \geq 0$,

$$K^\tau(x, y) = e^{-\lambda|x+y|} = e^{-\lambda(x+y)} = e^{-\lambda x} e^{-\lambda y}.$$

This factorization implies positive definiteness by Remark A.1.2.

(b) Here is a related example corresponding to a periodic function. Fix $\beta > 0, \lambda \geq 0$, and consider on $X = \mathbb{R}$ the β-periodic function given by

$$\varphi_\lambda(x) := e^{-\lambda x} + e^{-\lambda(\beta-x)} \quad \text{for} \quad 0 \leq x \leq \beta$$

(multiples of thermal euclidean Green's functions [DG13]). We claim that the kernel $K(x, y) := \varphi_\lambda(x - y)$ is reflection positive for $X_+ := [0, \beta/2]$.
   A direct calculation shows that the Fourier series of φ_λ is given by

$$\varphi_\lambda(x) = \sum_{n \in \mathbb{Z}} c_n e^{2\pi inx/\beta} \quad \text{with} \quad c_n = \frac{2\beta\lambda(1 - e^{-\beta\lambda})}{(\lambda\beta)^2 + (2\pi n)^2}. \quad (2.6)$$

As $c_n \geq 0$ for every $n \in \mathbb{Z}$, the function φ_λ is positive definite, i.e., K is positive definite.

Next we observe that, for $0 \leq x, y \leq \beta/2$, we have

$$K^\tau(x, y) = \varphi_\lambda(x + y) = e^{-\lambda(x+y)} + e^{-\lambda\beta}e^{\lambda(x+y)} = e^{-\lambda x}e^{-\lambda y} + e^{-\lambda\beta}e^{\lambda x}e^{\lambda y}.$$

Here both summands are positive definite kernels by Remark A.1.2.

Example 2.4.4 Reflection positive kernels show up naturally in the context of distributions if X is a manifold. We write $C_c^\infty(X)$ for the space of complex-valued compactly supported smooth functions on X and $C^{-\infty}(X)$ for the space of distributions, the space of continuous **anti-linear** functionals on $C_c^\infty(X) \to \mathbb{C}$ with respect to the natural LF topology on this space [Tr67].

The "distribution analog" of a positive definite kernel on X is a distribution $D \in C^{-\infty}(X \times X)$ which is *positive definite* in the sense the hermitian form

$$K_D(\varphi, \psi) := D(\varphi \otimes \overline{\psi}) =: \int_{X \times X} \overline{\varphi(x)}\psi(y)\, dD(x, y)$$

on $C_c^\infty(X)$ is positive semidefinite (this form is linear in the second argument because D is anti-linear). Then the corresponding reproducing kernel space $\mathscr{H}_D := \mathscr{H}_{K_D}$ consists of functions on $C_c(X)$ which are continuous and anti-linear, hence is a linear subspace of the space $C^{-\infty}(X)$ of distributions on X. The natural map

$$\iota_D \colon C_c^\infty(X) \to \mathscr{H}_D \subseteq C^{-\infty}(X), \quad \iota_D(\psi) = K_{D,\psi}, \quad \iota_D(\psi)(\varphi) = D(\varphi \otimes \overline{\psi}),$$

then has dense range and

$$\langle \iota_D(\varphi), \iota_D(\psi) \rangle = D(\varphi \otimes \overline{\psi}). \tag{2.7}$$

Definition 2.4.5 Let X be a smooth manifold, D a positive definite distribution on $X \times X$, let $\tau \colon X \to X$ be an involutive diffeomorphism of X and $X_+ \subseteq X$ be an open subset. We say that D is *reflection positive* with respect to (X, X_+, τ) if the distribution D^τ on $X_+ \times X_+$ defined by

$$D^\tau(\varphi) := D(\varphi \circ (\mathrm{id}_X \times \tau)) = \int_{X \times X} \overline{\varphi(x, \tau(y))}\, dD(x, y) \quad \text{for} \quad \varphi \in C_c^\infty(X_+ \times X_+)$$

is positive definite.

Specializing Lemma 2.4.2 to the context of reflection positive distributions, where the set X is replaced by the space $C_c^\infty(X)$, we obtain:

Lemma 2.4.6 *If the distribution D on $X \times X$ is reflection positive with respect to (X, X_+, τ), then $\mathscr{E} := \mathscr{H}_D$, $\theta(E)(\varphi) := E(\varphi \circ \tau)$ and $\mathscr{E}_+ := \overline{\iota_D(C_c^\infty(X_+))}$ defines a reflection positive Hilbert space of distributions. Further, $\widehat{\mathscr{E}} \cong \mathscr{H}_{D_+} \subseteq C^{-\infty}(X_+)$, where the map q is realized by*

$$q \colon \mathscr{E}_+ \to \widehat{\mathscr{E}} \cong \mathscr{H}_{D_+}, \quad q(E)(\varphi) := \langle \iota_D(\varphi), \theta E \rangle = E(\varphi \circ \tau).$$

Example 2.4.7 For $m > 0$, we consider the distribution $D_m := (m^2 - \Delta)^{-1}\delta_0$ on $\mathbb{R}^d$ which is the fundamental solution of the elliptic PDE $(m^2 - \Delta)D_m = \delta_0$. As $\delta_0 = \widehat{1}$ is the Fourier transform of the normalized Lebesgue measure $d\lambda_d(x) = \frac{dx}{(2\pi)^{d/2}}$ and $\frac{\partial}{\partial x_j}\widehat{D} = (-ix_j D)\widehat{\ }$ for any tempered distribution D on $\mathbb{R}^d$, it follows that $D_m = D_{\widehat{\nu_m}}$ for the measure

$$d\nu_m(p) = \frac{1}{m^2 + p^2}d\lambda_d(p) = \frac{1}{(2\pi)^{d/2}}\frac{dp}{m^2 + p^2}. \qquad (2.8)$$

For $d = 1$, we obtain a multiple of the function from Example 2.4.3:

$$\widehat{\nu_m}(x) = \frac{1}{\sqrt{2\pi}}\frac{\pi}{m}e^{-m|x|} = \frac{\sqrt{\pi}}{\sqrt{2m}}e^{-m|x|}.$$

It is easy to see that this distribution is reflection positive with respect to $(\mathbb{R}^d, \mathbb{R}^d_+, \tau)$, where

$$\mathbb{R}^d_+ = \{x = (x_0, \mathbf{x}): x_0 > 0\} \quad \text{and} \quad \tau(x_0, \mathbf{x}) = (-x_0, \mathbf{x})$$

is the reflection in the hyperplane $x_0 = 0$. First we observe that, for every test function ψ on $\mathbb{R}^d_+$, we have

$$D_m(\psi \otimes \theta\overline{\psi}) = \frac{1}{(2\pi)^{d/2}}\int_{\mathbb{R}^d} \overline{\widehat{\psi}(p_0, \mathbf{p})}\frac{\widehat{\psi}(-p_0, \mathbf{p})}{m^2 + p^2}\, dp \quad \text{for} \quad p = (p_0, \mathbf{p}) \in \mathbb{R} \times \mathbb{R}^{d-1}.$$

For each $\mathbf{p} \in \mathbb{R}^{d-1}$, the function $h_\mathbf{p}(p_0) := \widehat{\psi}(-p_0, \mathbf{p})$ is a Schwartz function with $\mathrm{supp}(\widehat{h}_\mathbf{p}) \subseteq (0, \infty)$, and

$$\int_{\mathbb{R}^d} \overline{\widehat{\psi}} \cdot \theta\widehat{\psi}\, d\nu_m = \int_{\mathbb{R}^{d-1}} \left(\int_{\mathbb{R}} \frac{\overline{h_\mathbf{p}(p_0)}h_\mathbf{p}(-p_0)}{p_0^2 + m^2 + \mathbf{p}^2}\, dp_0\right) d\mathbf{p}.$$

The reflection positivity of D_m now follows from $\int_{\mathbb{R}} \frac{\overline{h_\mathbf{p}(p_0)}h_\mathbf{p}(-p_0)}{p_0^2 + m^2 + \mathbf{p}^2}\, dp_0 \geq 0$ for $\mathbf{p} \in \mathbb{R}^{d-1}$, which is a consequence of Example 2.4.3a.

For $m = 0$, the measure $d\nu_0(p) = p^{-2}d\lambda_d(p)$ is locally finite if and only if $d \geq 3$. In this case the above arguments even show that $D_0 := D_{\widehat{\nu_0}}$ is a reflection positive distribution on $(\mathbb{R}^d, \mathbb{R}^d_+, \tau)$. For $d = 2$ we still obtain a reflection positive functional (defined in the obvious fashion) on the subspace of all test functions $\varphi \in C^\infty_c(\mathbb{R}^2)$ with $\int_{\mathbb{R}^2} \varphi(x)\, dx = 0$, and for $d = 1$ we have to impose in addition that $\int_{\mathbb{R}} x\varphi(x)\, dx = 0$.

In the following section we shall see a common geometric source of the preceding example and Example 2.4.3.

2.5 Reflection Positivity in Riemannian Geometry

In this section we describe a very natural class of reflection positive Hilbert spaces arising from isometric reflections of Riemannian manifolds.

Let M be a connected complete Riemannian manifold. An involution $\tau \in$ Isom(M) is called a *reflection* if there exists a fixed point $p \in M^\tau$ such that $T_p(\tau)$ is a hyperplane reflection in the tangent space $T_p(M)$. Then $\Sigma := M^\tau$ is a submanifold of M and the connected component containing p is of codimension one. We say that a reflection is *dissecting* if $M \setminus \Sigma$ has exactly two connected components which are exchanged by τ, i.e.,

$$M = M_+ \dot{\cup} \Sigma \dot{\cup} M_- \quad \text{with} \quad \tau(M_+) = M_-.$$

We consider the *Laplace–Beltrami operator* Δ_M on $L^2(M)$ as a negative self-adjoint operator on $L^2(M)$ [Str83, Theorem 2.4]. For each $m > 0$, we thus obtain a bounded positive operator $C := (m^2 - \Delta_M)^{-1}$ on $L^2(M)$.

Theorem 2.5.1 *If τ is a dissecting reflection on the connected complete Riemannian manifold M and $m > 0$. Then the involution θ on $L^2(M)$ defined by $\theta f := f \circ \tau$ satisfies*

$$\langle \varphi, C\theta\varphi \rangle \geq 0 \quad for \quad \varphi \in C_c^\infty(M_+).$$

Proof (cf. [An13, Theorem 8.3]) The starting point is the divergence formula on a Riemannian manifold M with boundary

$$\int_M \operatorname{div} X \, dV = \int_{\partial M} \langle X, \mathbf{n} \rangle \, dS, \tag{2.9}$$

where X is a compactly supported vector field and $\mathbf{n}$ is the outward normal vector field of ∂M.[1] In index notation, this reads

$$\int_M \nabla_a X^a \, dV = \int_{\partial M} n^a X_a \, dS. \tag{2.10}$$

For $\varphi \in C_c^\infty(M_+)$ and $u = C\varphi$ the function u is analytic in $M \setminus \operatorname{supp}(\varphi)$ because it satisfies the elliptic equation $(m^2 - \Delta)u = 0$ on this open subset. We now have

$$\langle C\varphi, \theta\varphi \rangle_{L^2} = \int_{M_-} \overline{C\varphi} \theta\varphi \, dV = \int_{M_-} \overline{u} C^{-1} \theta u \, dV$$

$$= \int_{M_-} \overline{u} C^{-1} \theta u - \theta(u) C^{-1} \overline{u} \, dV \quad (\varphi = C^{-1} u \text{ vanishes on } M_-)$$

$$= \int_{M_-} \overline{u}(m^2 - \Delta)\theta(u) - \theta(u)(m^2 - \Delta)\overline{u} \, dV = \int_{M_-} \theta(u)\Delta\overline{u} - \overline{u}\Delta\theta(u) \, dV.$$

---

[1] See [AF01, Sect. 3.8, Satz 26] and also [GHL87, Proposition 4.9], which has different sign conventions.

For $\Sigma = \partial M_-$, we also obtain

$$\int_\Sigma \theta(u)\nabla_{\mathbf{n}}\overline{u} - \overline{u}\nabla_{\mathbf{n}}\theta(u)\,dS = \int_\Sigma \langle \mathbf{n}, \theta(u)\nabla\overline{u} - \overline{u}\nabla\theta(u)\rangle\,dS$$

$$= \int_{M_-} \operatorname{div}\big(\theta(u)\nabla\overline{u} - \overline{u}\nabla\theta(u)\big)\,dV$$

$$= \int_{M_-} \langle \nabla\theta(u), \nabla\overline{u}\rangle + \theta(u)\Delta\overline{u} - \langle\nabla\overline{u}, \nabla\theta(u)\rangle - \overline{u}\Delta\theta(u)\,dV$$

$$= \int_{M_-} \theta(u)\Delta\overline{u} - \overline{u}\cdot\Delta\theta(u)\,dV.$$

This finally leads to

$$\langle C\varphi, \theta\varphi\rangle_{L^2} = \int_\Sigma \theta(u)\nabla_{\mathbf{n}}\overline{u} - \overline{u}\nabla_{\mathbf{n}}\theta(u)\,dS = \int_\Sigma \theta(u)\nabla_{\mathbf{n}}\overline{u} + \overline{u}\theta(\nabla_{\mathbf{n}}u)\,dS$$

$$= \int_\Sigma u\nabla_{\mathbf{n}}\overline{u} + \overline{u}\nabla_{\mathbf{n}}u\,dS = 2\operatorname{Re}\int_\Sigma \overline{u}\nabla_{\mathbf{n}}u\,dS.$$

Now

$$\int_\Sigma \overline{u}\nabla_{\mathbf{n}}u\,dS = \int_\Sigma \langle \mathbf{n}, \overline{u}\nabla u\rangle\,dS = \int_{M_-} \operatorname{div}(\overline{u}\nabla u)\,dV = \int_{M_-} \langle\nabla\overline{u}, \nabla u\rangle + \overline{u}\Delta u\,dV$$

$$= \int_{M_-} \langle\nabla\overline{u}, \nabla u\rangle + \overline{u}m^2 u\,dV = \|\nabla u\|_{L^2(M_-)}^2 + m^2\|u\|_{L^2(M_-)}^2$$

shows that $\langle C\varphi, \theta\varphi\rangle_{L^2} \geq 0$. $\qquad\square$

Remark 2.5.2 For $M = \mathbb{R}^d$ and $\tau(x_0, \mathbf{x}) = (-x_0, \mathbf{x})$, the reflection positivity of the distribution D_m in Example 2.4.7 is a very special case of Theorem 2.5.1.

Let $\mathscr{E} := \mathscr{H}^C$ denote the completion of $L^2(M)$ with respect to the scalar product $\langle f, h\rangle_C := \langle Cf, h\rangle_{L^2(M)}$. Then θ induces on $\mathscr{E}$ a unitary involution θ_C, and Theorem 2.5.1 implies that the subspace $\mathscr{E}_+$ generated by $C_c^\infty(M_+)$ is θ_C-positive. We thus obtain a reflection positive Hilbert space $(\mathscr{E}, \mathscr{E}_+, \theta_C)$.

Another interpretation of Theorem 2.5.1 is that the distribution D on $M \times M$ defined by $D(\varphi \otimes \psi) := \langle \varphi, C\overline{\psi}\rangle_{L^2(M)}$ is reflection positive with respect to (M, M_+, θ). From this perspective, we have $\mathscr{E} = \mathscr{H}_D$ as in Example 2.4.4 and $\widehat{\mathscr{E}}$ can be identified with the Hilbert space $\mathscr{H}_{D^\tau} \subseteq C^{-\infty}(M_+)$ of distributions on M_+.

2.6 Selfadjoint Extensions and Reflection Positivity

In this section we briefly indicate an operator theoretic approach to reflection positivity which makes it particularly clear how the space $\widehat{\mathscr{E}}$ depends on the choice of certain selfadjoint extensions of symmetric operators, resp., suitable boundary conditions.

We consider a Hilbert space $\mathcal{H}$ with a unitary involution θ and a closed subspace $\mathcal{H}_+$ such that $\mathcal{H}_- := \theta(\mathcal{H}_+) = \mathcal{H}_+^\perp$. Then we may identify $\mathcal{H}$ with $\mathcal{H}_+ \oplus \mathcal{H}_+$ on which θ acts by $\theta(v_+, v_-) = (v_-, v_+)$.

We consider a (densely defined) non-negative symmetric operator A on $\mathcal{D}_+ \subseteq \mathcal{H}_+$ and a selfadjoint extension L of A on $\mathcal{H}$ which commutes with θ and which is bounded from below.[2] For $-\lambda < \inf \operatorname{Spec}(L)$ we thus obtain a positive operator $\lambda \mathbf{1} + L$ with a bounded inverse

$$C := (\lambda \mathbf{1} + L)^{-1}.$$

Accordingly, we obtain on $\mathcal{H}$ a new scalar product $\langle v, w \rangle_C := \langle v, Cw \rangle$ and a corresponding completion $\mathcal{E} := \mathcal{H}^C$. We identify $\mathcal{H}$ with a linear subspace of $\mathcal{E}$ and write $\mathcal{E}_+$ for the closure of $\mathcal{H}_+$ in $\mathcal{E}$ and θ_C for the unitary involution on $\mathcal{E}$ obtained by extending θ.

Definition 2.6.1 We say that L is *reflection positive* if $(\mathcal{E}, \mathcal{E}_+, \theta_C)$ is a reflection positive Hilbert space, i.e., if

$$\langle \xi, \theta C \xi \rangle \geq 0 \quad \text{for} \quad \xi \in \mathcal{E}_+. \tag{2.11}$$

The following proposition shows that non-trivial spaces $\widehat{\mathcal{E}}$ can only be derived from operators L which are not simply the closure of $A \oplus \theta A \theta$ on $\mathcal{D}_+ \oplus \theta(\mathcal{D}_+)$.

Proposition 2.6.2 *If the symmetric operator A is essentially selfadjoint on $\mathcal{H}_+$, then L is reflection positive and $\widehat{\mathcal{E}} = \{0\}$.*

Proof As $\lambda \mathbf{1} + L$ is strictly positive, there exists an $\varepsilon > 0$ with

$$\langle (L + \lambda \mathbf{1})\xi, \xi \rangle \geq \varepsilon \|\xi\|^2 \quad \text{for} \quad \xi \in \mathcal{D}(L).$$

This implies in particular that $\langle (A + \lambda \mathbf{1})\xi, \xi \rangle \geq \varepsilon \|\xi\|^2$ for $\xi \in \mathcal{D}_+$. Since A is essentially selfadjoint and non-negative, it follows that the selfadjoint operator $\lambda \mathbf{1} + \overline{A}$ on $\mathcal{E}_+$ satisfies $\lambda \mathbf{1} + \overline{A} \geq \varepsilon$. In particular, it is invertible on $\mathcal{H}_+$. Therefore $\mathcal{R}(\lambda \mathbf{1} + A) = (\lambda \mathbf{1} + A)\mathcal{D}_+$ is dense in $\mathcal{H}_+$. We conclude that the continuous operator $C = (\lambda \mathbf{1} + L)^{-1}$ maps the dense subspace $\mathcal{R}(\lambda \mathbf{1} + A)$ of $\mathcal{H}_+$ into $\mathcal{H}_+$, so that $C\mathcal{H}_+ \subseteq \mathcal{H}_+$. Now $\theta C \theta = C$ further implies that $C\mathcal{H}_- \subseteq \mathcal{H}_-$, so that $\langle \theta \xi, \xi \rangle_C = 0$ for $\xi \in \mathcal{H}_+$. This shows that L is reflection positive with $\widehat{\mathcal{E}} = \{0\}$. □

Corollary 2.6.3 *If $\mathcal{H}$ is finite dimensional, then $\widehat{\mathcal{E}} = \{0\}$.*

If L is reflection positive, then the continuous linear map $q_{\mathcal{H}} := q|_{\mathcal{H}_+} : \mathcal{H}_+ \to \widehat{\mathcal{E}}$ has dense range, so that its adjoint $q_{\mathcal{H}}^* : \widehat{\mathcal{E}} \to \mathcal{H}_+$ is injective. We may therefore consider $\widehat{\mathcal{E}}$ as a linear subspace of $\mathcal{H}_+$. The following observation shows that the image of $\widehat{\mathcal{E}}$ in $\mathcal{H}_+$ consists of solutions of the eigenvalue equation

---

[2]See [AS80, AG82] for a systematic discussion of the set of positive extensions of positive symmetric operators.

$$A^*\xi = -\lambda\xi, \quad \xi \in \mathcal{D}(A^*) \subseteq \mathcal{H}_+.$$

Proposition 2.6.4 *Suppose that L is reflection positive. Then $q_{\mathcal{H}}$ maps the eigenspace $\ker(\lambda\mathbf{1} + A^*)$ onto a dense subspace of $\widehat{\mathcal{E}}$, the image of $q_{\mathcal{H}}^*$ is contained in $\ker(\lambda\mathbf{1} + A^*)$, and*

$$\ker(q_{\mathcal{H}}) = (\lambda\mathbf{1} + L)(\mathcal{D}(L) \cap \mathcal{H}_+) \cap \mathcal{H}_+. \tag{2.12}$$

Proof For $\xi \in \mathcal{H}_+$, the relation $q_{\mathcal{H}}(\xi) = 0$ is equivalent to

$$0 = \langle \eta, C\theta(\xi) \rangle = \langle (\lambda\mathbf{1} + L)^{-1}\eta, \theta(\xi) \rangle \quad \text{for all} \quad \eta \in \mathcal{H}_+. \tag{2.13}$$

If $\xi = (\lambda\mathbf{1} + A)\zeta$ for $\zeta \in \mathcal{D}_+$, then also $\xi = (\lambda\mathbf{1} + L)\zeta$, so that

$$C\theta(\xi) = \theta C\xi = \theta\zeta \in \mathcal{H}_-$$

implies that $\mathcal{R}(\lambda\mathbf{1} + A) = (\lambda\mathbf{1} + A)\mathcal{D}_+ \subseteq \ker(q_{\mathcal{H}})$. We now obtain

$$\mathrm{im}(q_{\mathcal{H}}^*) \subseteq \ker(q_{\mathcal{H}})^\perp \subseteq \mathcal{R}(\lambda\mathbf{1} + A)^\perp = \ker(\lambda\mathbf{1} + A^*).$$

This in turn shows that the restriction of $q_{\mathcal{H}}$ to $\mathcal{R}(\lambda\mathbf{1} + A)^\perp = \ker(\lambda\mathbf{1} + A^*)$ has dense range. Finally, we note that (2.13) is equivalent to $C\theta\xi = \theta C\xi \in \mathcal{H}_+^\perp = \mathcal{H}_-$, which in turn is equivalent to $C\xi \in \mathcal{H}_+$, i.e., to $\xi \in (\lambda\mathbf{1} + L)\mathcal{H}_+$. This proves (2.12). □

For general results on the existence of reflection positive extensions of semi-bounded symmetric operators, we refer to [Ne18].

Example 2.6.5 The preceding discussion is an operator theoretic abstraction of the geometric example in Sect. 2.5. To match the abstract framework, we put $\mathcal{H} := L^2(M)$, $\mathcal{H}_\pm := L^2(M_\pm)$ and consider the positive selfadjoint operator $L := -\Delta$ as a θ-invariant extension of the restriction $A := -\Delta|_{C_c^\infty(M_+)}$. In this case the eigenvalue equation

$$A^* f = -\lambda f \quad \text{for} \quad f \in \mathcal{D}(A^*) \subseteq \mathcal{H}_+$$

is equivalent to $(\lambda\mathbf{1} - \Delta)f \perp C_c^\infty(M_+)$, which means that $f \in \mathcal{H}_+ = L^2(M_+)$ satisfies the PDE

$$\Delta f = \lambda f \quad \text{on} \quad M_+ \tag{2.14}$$

in the distribution sense. Ellipticity of Δ implies that f can be represented on M_+ by an analytic function ([Ru73, Theorem 8.12]). We thus obtain a realization of $\widehat{\mathcal{E}}$ in the space of L^2-solutions of (2.14) on the open subset M_+.

Example 2.6.6 For the simple example $M = \mathbb{R}$ with $\tau(x) = -x$ and $M_+ = (0, \infty)$ with $Af = -f''$, we consider $\lambda = m^2$ for some $m > 0$. Then the solutions of (2.14) on $\mathbb{R}_+$ are for $\lambda = m^2$ of the form $f(x) = ae^{mx} + be^{-mx}$, so that the L^2-condition leads to $f(x) = be^{-mx}$. This already shows that $\dim \widehat{\mathscr{E}} \leq 1$ and that $q_{\mathscr{H}}$ must be a multiple of the linear functional $h \mapsto \int_0^\infty h(x)e^{-mx}\,dx = \mathscr{L}(h)(m)$ (cf. Example 2.4.3).

Notes

The Markov condition (2.3) in Sect. 2.3 is an abstraction of the Markov condition for Osterwalder–Schrader positive processes that one finds in [Kl77, Kl78, NÓ15a]. For a detailed analysis of its operator theoretic aspects we refer to [JT17], where one also finds a discussion of reflection positive Hilbert spaces in terms of graphs of contractions.

Example 2.4.7 corresponds to OS-positivity for free fields in d-space ([GJ81, Ja08]).

Theorem 2.5.1 and several variants can be found in [An13] and the work of Jaffe and Ritter [JR07]; see also [AFG86, Theorem 2] and [Di04, Theorem 2] for related results.

Chapter 3
Reflection Positive Representations

In this chapter we turn to operators on reflection positive (real or complex) Hilbert spaces and introduce the Osterwalder–Schrader transform to pass from operators on $\mathscr{E}_+$ to operators on $\widehat{\mathscr{E}}$ (Sect. 3.1). The objects represented in reflection positive Hilbert spaces $(\mathscr{E}, \mathscr{E}_+, \theta)$ are symmetric Lie groups (G, τ), i.e., a Lie group G, endowed with an involutive automorphism τ. A typical example in physics arises from the euclidean motion group and time reversal. There are several ways to specify compatibility of a unitary representation $(U, \mathscr{E})$ of (G, τ) with $\mathscr{E}_+$ and θ and thus to define reflection positive representations (Sect. 3.3). One is to specify a subset $G_+ \subseteq G$ and assume that $\mathscr{E}_+$ is generated by applying G_+^{-1} to a suitable subspace of $\mathscr{E}_+$. The other simpler one applies if $S := G_+^{-1}$ is a subsemigroup of G invariant under the involution $s \mapsto s^\sharp = \tau(s)^{-1}$. Then we simply require $\mathscr{E}_+$ to be S-invariant. In both cases we can use the integrability results in Chap. 7 to obtain unitary representations of the 1-connected Lie group G^c with Lie algebra $\mathfrak{g}^c = \mathfrak{h} + i\mathfrak{q}$ on $\widehat{\mathscr{E}}$. As reflection positive unitary representations are mostly constructed by applying a suitable Gelfand–Naimark–Segal (GNS) construction to reflection positive functions, we discuss this correspondence in some detail in Sect. 3.4. In particular, we discuss the Markov condition in this context (Proposition 3.4.9).

3.1 The OS Transform of Linear Operators

We have already seen how to pass from a reflection positive Hilbert space $(\mathscr{E}, \mathscr{E}_+, \theta)$ to the new Hilbert space $\widehat{\mathscr{E}}$. We now follow this passage for linear operators on $\mathscr{E}_+$.

Definition 3.1.1 (*OS transform*) Suppose that $S \colon \mathscr{E}_+ \supseteq \mathscr{D}(S) \to \mathscr{E}_+$ is a linear operator (not necessarily bounded) with $S(\mathscr{D}(S) \cap \mathscr{N}) \subseteq \mathscr{N}$. Then S induces a linear operator

$$\widehat{S} \colon \mathscr{D}(\widehat{S}) := \widehat{\mathscr{D}(S)} = \{\widehat{v} \colon v \in \mathscr{D}(S)\} \to \widehat{\mathscr{E}}, \quad \widehat{S}\widehat{\eta} := \widehat{S\eta}.$$

© The Author(s) 2018
K.-H. Neeb and G. Ólafsson, *Reflection Positivity*, SpringerBriefs in Mathematical Physics, https://doi.org/10.1007/978-3-319-94755-6_3

The passage from S to $\widehat{S}$ is called the *Osterwalder–Schrader transform* (or OS transform for short).

Lemma 3.1.2 *Let $(\mathscr{E}, \mathscr{E}_+, \theta)$ be a real of complex reflection positive Hilbert space. Suppose that $\mathscr{D} \subseteq \mathscr{E}_+$ is a linear subspace such that $\widehat{\mathscr{D}} = \{\widehat{v} \colon v \in \mathscr{D}\}$ is dense in $\widehat{\mathscr{E}}$, and that $S, T \colon \mathscr{D} \to \mathscr{E}_+$ are linear operators. Then the following assertions hold:*

(a) *If $\langle S\eta, \zeta \rangle_\theta = \langle \eta, T\zeta \rangle_\theta$ for $\eta, \zeta \in \mathscr{D}$, then $S(\mathscr{N}) \subseteq \mathscr{N}$, so that $\widehat{S}, \widehat{T} \colon \widehat{\mathscr{D}} \to \widehat{\mathscr{E}}$ are well-defined and*

$$\langle \widehat{S}\widehat{\eta}, \widehat{\zeta} \rangle = \langle \widehat{\eta}, \widehat{T}\widehat{\zeta} \rangle \quad \textit{for} \quad \widehat{\eta}, \widehat{\zeta} \in \widehat{\mathscr{D}}.$$

(b) *Let $\widetilde{S} \in \mathrm{U}(\mathscr{E})$ be unitary with $\widetilde{S}\mathscr{E}_+ = \mathscr{E}_+$ and $\theta\widetilde{S}\theta = \widetilde{S}$. For $\mathscr{D}_+ = \mathscr{E}_+$ and $S := \widetilde{S}|_{\mathscr{E}_+}$, the operator $\widehat{S}$ extends to a unitary operator on $\widehat{\mathscr{E}}$.*

(c) *If $\langle S\eta, \zeta \rangle_\theta = \langle \eta, S\zeta \rangle_\theta$ for all $\eta, \zeta \in \mathscr{D}$, then $\widehat{S}$ is a symmetric operator. If, in addition, S is bounded and $\mathscr{D} = \mathscr{E}_+$, then so is $\widehat{S}$, and $\|\widehat{S}\| \leq \|S\|$.*

(d) *If $U \in \mathrm{U}(\mathscr{E})$ satisfies $U\mathscr{E}_+ = \mathscr{E}_+$ and $\theta U\theta = U^{-1}$, then $\widehat{U}^2 = \mathrm{id}_{\widehat{\mathscr{E}}}$. Further, $\mathscr{E}$ is a direct sum of reflection positive Hilbert subspaces $(\mathscr{F}, \mathscr{F} \cap \mathscr{E}_+, \theta|_{\mathscr{F}})$ and $(\mathscr{G}, \mathscr{G} \cap \mathscr{E}_+, \theta|_{\mathscr{G}})$, invariant under U and U^{-1}, such that $\widehat{\mathscr{G}} = \{0\}$ and $(U|_{\mathscr{F}})^2 = \mathbf{1}$.*

Proof (a) For $\eta, \zeta \in \mathscr{D}$, we obtain from $\langle S\eta, \zeta \rangle_\theta = \langle \eta, T\zeta \rangle_\theta$ that $\eta \in \mathscr{N}$ implies that $\widehat{S}\eta = 0$, i.e., $S\eta \in \mathscr{N}$. Therefore $\widehat{S}\widehat{\eta} := \widehat{S\eta}$ is well-defined and the remainder of (a) follows.

(b) In this case (a) holds with $T = S^{-1}$, so that $\widehat{S}$ and $\widehat{T}$ are well-defined and mutually inverse on $\widehat{\mathscr{D}}$. In particular, we have $S\widehat{\mathscr{D}} = \widehat{\mathscr{D}}$. From

$$\langle S\eta, S\zeta \rangle_\theta = \langle S\eta, \theta S\zeta \rangle = \langle \widetilde{S}\eta, \widetilde{S}\theta\zeta \rangle = \langle \eta, \zeta \rangle_\theta \quad \textit{for} \quad \xi, \eta \in \mathscr{E}_+,$$

it further follows that $\widehat{S} \colon \widehat{\mathscr{D}} \to \widehat{\mathscr{D}}$ is unitary. Therefore it extends uniquely to a unitary operator on $\widehat{\mathscr{E}}$.

(c) From (a) it follows that $\widehat{S}$ is well-defined and symmetric. Now we assume that S is bounded and defined on all of $\mathscr{E}_+$. Then

$$\|\widehat{S}^k\widehat{\eta}\|^2 = \langle \widehat{\eta}, \widehat{S}^{2k}\widehat{\eta} \rangle \leq \|\widehat{\eta}\| \|\widehat{S}^{2k}\widehat{\eta}\| \quad \textit{for} \quad \eta \in \mathscr{E}_+$$

and therefore

$$\|\widehat{S}\widehat{\eta}\|^{2^n} \leq \|\widehat{S}^2\widehat{\eta}\|^{2^{n-1}} \|\widehat{\eta}\|^{2^{n-1}} \leq \|\widehat{S}^4\widehat{\eta}\|^{2^{n-2}} \|\widehat{\eta}\|^{2^{n-1}+2^{n-2}} \leq \cdots \leq \|\widehat{S}^{2^n}\widehat{\eta}\| \|\widehat{\eta}\|^{2^n-1}.$$

We also have $\|\widehat{S}^m\widehat{\eta}\|^2 = \langle \theta S^m\eta, S^m\eta \rangle \leq \|S\|^{2m} \|\eta\|^2$, which leads to

$$\|\widehat{S}\widehat{\eta}\|^{2^n} \leq \|S\|^{2^n} \|\eta\| \|\widehat{\eta}\|^{2^n-1}.$$

We conclude that

$$\|\widehat{S}\widehat{\eta}\| \leq \|S\| \cdot \lim_{n \to \infty} \left(\|\eta\|^{2^{-n}} \|\widehat{\eta}\|^{1-2^{-n}} \right) = \|S\| \|\widehat{\eta}\|.$$

Therefore $\widehat{S}$ is bounded with $\|\widehat{S}\| \leq \|S\|$.

(d) From (c) it follows that $\widehat{U}$ is a well-defined symmetric contraction. The same argument applies to $V := U^{-1}$ and leads to another symmetric contraction $\widehat{V}$. Now $\widehat{U}\widehat{V}\widehat{v} = \widehat{UV}\widehat{v} = \widehat{v}$ for every $v \in \mathscr{E}_+$ implies that $\widehat{U}\widehat{V} = \mathrm{id}_{\widehat{\mathscr{E}}}$. We likewise get $\widehat{V}\widehat{U} = \mathrm{id}_{\widehat{\mathscr{E}}}$, so that $\widehat{U}^{-1} = \widehat{V}$. This shows that $\widehat{U}^{-1}$ also is a symmetric contraction. We conclude that $(\widehat{U}) \subseteq \{-1, 1\}$, which further leads to $\widehat{U}^2 = \mathrm{id}_{\widehat{\mathscr{E}}}$.

Next we observe that $\mathscr{E}_+$ is invariant under U and U^{-1}, so that $\mathscr{E}^0 := \overline{\mathscr{E}_+ + \theta(\mathscr{E}_+)}^{\perp}$ is also invariant under $U^{\pm 1}$ and θ. Since the closed subspace $\mathscr{N} \subseteq \mathscr{E}_+$ is invariant under U and $V = U^{-1}$, the subspace $\mathscr{E}^1 := \mathscr{N} \oplus \theta(\mathscr{N}) \subseteq (\mathscr{E}^0)^{\perp}$ is invariant under U, U^{-1} and θ, and this property is inherited by $\mathscr{E}^2 := (\mathscr{E}^0 \oplus \mathscr{E}^1)^{\perp}$. With $\mathscr{E}_+^j := \mathscr{E}^j \cap \mathscr{E}_+$, we now obtain a direct sum decomposition of the reflection positive Hilbert space $(\mathscr{E}, \mathscr{E}_+, \theta)$ into the orthogonal sum of the three reflection positive spaces $(\mathscr{E}^j, \mathscr{E}_+^j, \theta|_{\mathscr{E}^j})$, $j = 0, 1, 2$. We put $\mathscr{G} = \mathscr{E}^0 \oplus \mathscr{E}^1$ and $\mathscr{F} := \mathscr{E}^2$. As $\mathscr{E}_+^0 = \{0\}$ and $\mathscr{E}_+^1 = \mathscr{N}$, we have $\widehat{\mathscr{G}} = \{0\}$ and, accordingly, $\widehat{\mathscr{F}} = \widehat{\mathscr{E}}$. Further $\mathscr{N} \cap \mathscr{F}_+ = \{0\}$ implies that $q|_{\mathscr{F}_+}$ is injective. Hence $q \circ U|_{\mathscr{E}_+} = \widehat{U} \circ q$ implies that $U_+ := U|_{\mathscr{F}_+}$ also satisfies $U_+^2 = 1$. Likewise $U|_{\theta\mathscr{F}_+} = \theta U_+ \theta$ is an involution. By construction, $\mathscr{F}_+ + \theta(\mathscr{F}_+)$ is dense in $\mathscr{F}$, and this leads to $(U|_{\mathscr{F}})^2 = 1$. $\qquad\square$

Remark 3.1.3 (a) Typical operators to which part (b) of the preceding lemma applies are unitary operators $S \in \mathrm{U}(\mathscr{E})$ with $S\mathscr{E}_+ = \mathscr{E}_+$ and $\theta S \theta = S$.

(b) Suppose that $\mathscr{E}$ is finite-dimensional and that $U \in \mathrm{U}(\mathscr{E})$ satisfies $U\mathscr{E}_+ \subseteq \mathscr{E}_+$ and $\theta U \theta = U^{-1}$. Then the finite dimension implies that $U\mathscr{E}_+ = \mathscr{E}_+$, so that Lemma 3.1.2(d) shows that $\widehat{U}^2 = 1$.

For symmetries of the whole structure encoded in $(\mathscr{E}, \mathscr{E}_+, \theta)$, the corresponding actions on $\mathscr{E}$, resp., $\mathscr{E}_+$ lead to unitary operators on $\widehat{\mathscr{E}}$:

Proposition 3.1.4 *Let $\mathscr{E}$ be a real or complex Hilbert space, θ be a unitary involution on $\mathscr{E}$, and $\mathscr{E}_+ \subseteq \mathscr{E}$ be a θ-positive subspace. Suppose that $(U, \mathscr{E})$ is a strongly continuous unitary representation of a topological group G on $\mathscr{E}$ such that*

$$U_g \mathscr{E}_+ \subseteq \mathscr{E}_+ \quad and \quad U_g \theta = \theta U_g \quad for \quad g \in G.$$

Then the OS transform defines a continuous unitary representation $(\widehat{U}, \widehat{\mathscr{E}})$ of G.

As we shall see below, far more interesting situations arise from unitary representations not commuting with θ and not leaving $\mathscr{E}_+$ invariant. The structure required in this context is introduced in the following section.

3.2 Symmetric Lie Groups and Semigroups

Definition 3.2.1 (*Symmetric Lie groups*) Let G be a Lie group with Lie algebra $\mathfrak{g}$ and let $\tau : G \to G$ be an involutive automorphism. We then call (G, τ) a *symmetric Lie group*. Likewise, a *symmetric Lie algebra* $(\mathfrak{g}, \tau)$ is a Lie algebra $\mathfrak{g}$, endowed with an involutive automorphism of $\mathfrak{g}$.

We shall see below that it is often convenient to encode τ in the larger group

$$G_\tau := G \rtimes \{\mathrm{id}_G, \tau\}. \qquad (3.1)$$

Then $\tau \in G_\tau$ and conjugation with τ on the normal subgroup G satisfies $\tau g \tau = \tau(g)$ for $g \in G$.

We put $H := (G^\tau)_0$, where $_0$ stands for the connected component containing the identity element e.

The involution τ induces an involution $\mathrm{d}\tau : \mathfrak{g} \to \mathfrak{g}$. We also write

$$\mathfrak{h} := \{x \in \mathfrak{g} : \mathrm{d}\tau(x) = x\} \quad \text{and} \quad \mathfrak{q} := \{x \in \mathfrak{g} : \mathrm{d}\tau(x) = -x\}.$$

Then $\mathfrak{g} = \mathfrak{h} \oplus \mathfrak{q}$ and $\mathfrak{h}$ is the Lie algebra of H. Furthermore,

$$[\mathfrak{h}, \mathfrak{h}] + [\mathfrak{q}, \mathfrak{q}] \subseteq \mathfrak{h} \quad \text{and} \quad [\mathfrak{h}, \mathfrak{q}] \subseteq \mathfrak{q}.$$

In particular $\mathfrak{g}^c := \mathfrak{h} \oplus i\mathfrak{q}$ is a Lie subalgebra of the complexification $\mathfrak{g}_{\mathbb{C}} = \mathfrak{g} + i\mathfrak{g}$, called the *Cartan dual* of $\mathfrak{g}$. We denote by G^c a simply connected Lie group with Lie algebra $\mathfrak{g}^c$. We observe that

$$g^\sharp := \tau(g)^{-1} \quad \text{satisfies} \quad (g^\sharp)^\sharp = g \quad \text{and} \quad (gh)^\sharp = h^\sharp g^\sharp, \qquad (3.2)$$

so that $\sharp$ defines on G the structure of an involutive (semi-) group.

Example 3.2.2 Let $G = E(n) = \mathbb{R}^n \rtimes O_n(\mathbb{R})$ be the euclidean motion group and $\mathfrak{g} = \mathfrak{e}(n)$ be its Lie algebra. Its elements (b, A) act on $\mathbb{R}^n$ by $(b, A).v = Av + b$. The product in G is given by $(x, A)(y, B) = (x + Ay, AB)$.

Let $r_0 := \mathrm{diag}(-1, 1, \ldots, 1)$ and define an involution on G by

$$\tau(x, A) = (r_0 x, r_0 A r_0).$$

As

$$r_0 \begin{pmatrix} a & b \\ c & D \end{pmatrix} r_0 = \begin{pmatrix} a & -b \\ -c & D \end{pmatrix} \quad \text{for} \quad \begin{pmatrix} a & b \\ c & D \end{pmatrix} \in M_n(\mathbb{R}),$$

$\mathfrak{g}^c \simeq (i\mathbb{R} \times \mathbb{R}^{n-1}) \rtimes \mathfrak{so}_{1,n-1}(\mathbb{R}) \simeq \mathbb{R}^{1,n-1} \rtimes \mathfrak{so}_{1,n-1}(\mathbb{R}) =: \mathfrak{p}(n)$ is the Lie algebra of the Poincaré group $P(n)$. We then obtain the duality relation

$$\mathfrak{e}(n)^c \cong \mathfrak{p}(n),$$

which is of fundamental importance in physics (cf. Chap. 8).

Example 3.2.3 (a) The affine group $\mathrm{Aff}(V) \cong V \rtimes \mathrm{GL}(V)$ of a vector space V carries a natural structure of a symmetric Lie group. We write its elements as pairs (b, A), corresponding to the map $v \mapsto Av + b$. Then $\tau(b, A) := (-b, A)$ is an involutive automorphism of $\mathrm{Aut}(V)$.

For $G = \mathrm{Aff}(V)$, we then have $G^\tau = \mathrm{GL}(V)$, $\mathfrak{g} \cong V \rtimes \mathfrak{gl}(V)$, $\mathfrak{h} = \mathfrak{gl}(V)$ and $\mathfrak{q} \cong V$. Since $[\mathfrak{q}, \mathfrak{q}] = \{0\}$, we have $\mathfrak{g}^c \cong \mathfrak{g}$.

(b) We obtain a particularly important example for $V = \mathbb{R}$, the "*ax + b*-group". Here

$$S := (\mathbb{R}_{\geq 0}, +) \rtimes (\mathbb{R}_+^\times, \cdot) = \{(b, a) : (b, a)\mathbb{R}_+ \subseteq \mathbb{R}_+\} = \{(b, a) : b \geq 0, a > 0\}$$

is a closed $\sharp$-invariant subsemigroup of $\mathrm{Aff}(\mathbb{R})$.

Example 3.2.4 If $r \in O_n(\mathbb{R})$ is any involution of determinant -1, then $\tau(g) := rgr$ defines an involutive automorphism of $\mathrm{SO}_n(\mathbb{R})$ such that $O_n(\mathbb{R}) \cong \mathrm{SO}_n(\mathbb{R})_\tau$ in the sense of (3.1).

Definition 3.2.5 A *symmetric semigroup* is a triple (G, S, τ), where (G, τ) is a symmetric Lie group and $S \subseteq G$ is a subsemigroup satisfying

(S1) S is invariant under $s \mapsto s^\sharp$, so that $(S, \sharp)$ is an involutive semigroup.
(S2) $HS = S$.
(S3) $\mathbf{1} \in \overline{S}$.

If (S1) holds for a subsemigroup $S \subseteq G$ we simply call it a *symmetric subsemigroup of* (G, τ). We shall mostly use only (S1). Note that (S1/2) imply that also $SH = (HS)^\sharp = S$.

Examples 3.2.6 (a) $(\mathbb{R}, \mathbb{R}_+, -\mathrm{id}_\mathbb{R})$ and $(\mathbb{Z}, \mathbb{N}_0, -\mathrm{id}_\mathbb{Z})$ are the most elementary examples of symmetric semigroups.

(b) If $\mathbb{R}^{1,d-1} \cong \mathbb{R}^d$ is d-dimensional Minkowski space and $G = (\mathbb{R}^d, +)$ its translation group, then time reversal $\tau(x_0, \mathbf{x}) = (-x_0, \mathbf{x})$ is an involutive automorphism and the open light cone $V_+ \subseteq \mathbb{R}^{1,d-1}$ is a subsemigroup invariant under the map $x \mapsto x^\sharp = -\tau(x) = (x_0, -\mathbf{x})$.

A closely related example is the euclidean space $G = (\mathbb{R}^d, +)$ with the same involution and the open half space $S = \mathbb{R}_+^d = \{(x_0, \mathbf{x}) : x_0 > 0\}$.

(c) Semigroups with polar decomposition: Let (G, τ) be a symmetric Lie group and H be an open subgroup of $G^\tau = \{g \in G : \tau(g) = g\}$. We denote the derived involution $\mathfrak{g} \to \mathfrak{g}$ by the same letter and define $\mathfrak{h} = \{x \in \mathfrak{g} : \tau(x) = x\} = \mathfrak{g}^\tau$ and $\mathfrak{q} = \{x \in \mathfrak{g} : \tau(x) = -x\} = \mathfrak{g}^{-\tau}$. Then $\mathfrak{g} = \mathfrak{h} \oplus \mathfrak{q}$. We say that the open subsemigroup $S \subseteq G$ has a polar decomposition if there exists an H-invariant open convex cone $C \subset \mathfrak{q}$ such that $S = H \exp C$ and the map $H \times C \to S$, $(h, X) \mapsto h \exp X$ is a diffeomorphism (cf. [La94, Nel64, HN93]). Typical examples are the complex Olshanski semigroups in complex simple Lie groups such as $\mathrm{SU}_{p,q}(\mathbb{C})_\mathbb{C} \cong$

$\mathrm{SL}_{p+q}(\mathbb{C})$. Complex Olshanski semigroups exist if and only if the non-compact Riemannian symmetric space associated to G is a bounded symmetric domain. This is equivalent to the existence of a G-invariant convex cone $C \subset i\mathfrak{g}$ such that $G \exp C$ is a subsemigroup of $G_{\mathbb{C}}$, if $G_{\mathbb{C}} \supseteq G$ is an injective complexification of G. More generally, we have the causal symmetric spaces of non-compact type like *de Sitter space* $\mathrm{dS}^n \cong \mathrm{SO}_{1,n}(\mathbb{R})^\uparrow / \mathrm{SO}_{1,n-1}(\mathbb{R})^\uparrow$ ([HÓ97]; see also Sect. 7.3.3. In this case $\mathfrak{q} \simeq \mathbb{R}^{1,n-1}$ is the n-dimensional Minkowski space and C corresponds to the open light-cone in $\mathfrak{q}$.

(d) The simply connected covering group $G := \tilde{\mathrm{SL}}_2(\mathbb{R})$ of $\mathrm{SL}_2(\mathbb{R})$ carries an involution τ acting on $\mathfrak{g} = \mathfrak{sl}_2(\mathbb{R})$ by

$$\tau \begin{pmatrix} x & y \\ z & -x \end{pmatrix} = \begin{pmatrix} x & -y \\ -z & -x \end{pmatrix},$$

and there exists a closed subsemigroup $S \subseteq G$ whose boundary is

$$\partial S = H(S) := S \cap S^{-1} = \exp(\mathfrak{b}) \quad \text{with} \quad \mathfrak{b} := \left\{ \begin{pmatrix} x & y \\ 0 & -x \end{pmatrix} : x, y \in \mathbb{R} \right\}.$$

This semigroup satisfies $S^\sharp = S$, and the subgroup $H(S)$ is τ-invariant, but strictly larger than G_0^τ.

3.3 Reflection Positive Representations

Suppose that (G, τ) is a symmetric Lie group. For a unitary representation $(U, \mathscr{E})$ of G on the reflection positive Hilbert space $(\mathscr{E}, \mathscr{E}_+, \theta)$, the condition $\theta U_g \theta = U_{\tau(g)}$ for $g \in G$ is equivalent to $U_\tau := \theta$ defining a unitary representation $U : G_\tau \to \mathrm{U}(\mathscr{E})$ (cf. (3.1)). Accordingly, we shall always work with representations of the enlarged group G_τ in the following and assume that $\theta = U_\tau$.

Next we address the additional requirements that make a unitary representation $(U, \mathscr{H})$ of G_τ on a reflection positive Hilbert space compatible with the subspace $\mathscr{E}_+$. An obvious natural assumption is that the operators $(U_h)_{h \in H}$ act by automorphisms of the full structure, i.e., $U_h \mathscr{E}_+ = \mathscr{E}_+$ for $h \in H$. Since U_H commutes with θ, it preserves both eigenspaces $\mathscr{E}_{\pm 1} = \ker(\theta \mp \mathbf{1})$. If $\mathscr{E}_+ = \mathscr{G}(C)$ is the graph of a contraction $C : \mathscr{E}_1 \supseteq \mathscr{F} \to \mathscr{E}_{-1}$ as in Sect. 2.2, then the invariance of $\mathscr{E}_+$ under U_H is equivalent to C being an intertwining operator for the representations of H on $\mathscr{E}_{+1}$ and $\mathscr{E}_{-1}$.

Eventually, one would like to impose conditions that can be used to derive a unitary representation of the simply connected Lie group G^c with Lie algebra $\mathfrak{g}^c$ on the space $\hat{\mathscr{E}}$. The group G^c always contains a subgroup with the Lie algebra $\mathfrak{h}$, so that the representation of this subgroup is provided directly by Proposition 3.1.4, but for the operators generated by the subspace $i\mathfrak{q} \subseteq \mathfrak{g}^c$ it is less clear how they should be obtained (cf. Chap. 7).

One way to express such requirements uses to symmetric subsemigroups of G, but in many relevant examples there are no such subsemigroups with interior points and one has to consider more general domains $G_+ \subseteq G$.

Definition 3.3.1 Let (G, τ) be a symmetric Lie group and $(U, \mathscr{E})$ be a unitary representation of G_τ on the Hilbert space $\mathscr{E}$. We put $\theta := U_\tau$.

(a) Let $G_+ \subseteq G$ be a subset. We consider a real linear space V and a linear map $j: V \to \mathscr{H}$ whose range is cyclic for the unitary representation U of G_τ, i.e., $[\![U_{G_\tau} j(V)]\!] = \mathscr{H}$. Then we say that $(U, \mathscr{E}, j, V)$ is *reflection positive with respect to the subset* $G_+ \subseteq G$ if the subspace $\mathscr{E}_+ := [\![U_{G_+}^{-1} j(V)]\!]$ is θ-positive.

(b) If $S \subseteq G$ is a $\sharp$-invariant subsemigroup and $(\mathscr{E}, \mathscr{E}_+, \theta)$ is a reflection positive Hilbert space, then $(U, \mathscr{E})$ is said to be *reflection positive with respect to S* if $U_s \mathscr{E}_+ \subseteq \mathscr{E}_+$ for every $s \in S$. Then the conditions under (a) are satisfied for $V = \mathscr{E}_+$, $j = \mathrm{id}_V$, and $G_+ := S^{-1} = \tau(S)$.

Lemma 3.3.2 *Let* $(\mathscr{E}, \mathscr{E}_+, \theta)$ *be a reflection positive Hilbert,* $\mathscr{E}_- := \theta \mathscr{E}_+$ *and put* $U^\sharp := \theta U^* \theta$ *for* $U \in \mathrm{U}(\mathscr{E})$. *Then*

$$S(\mathscr{E}_+) := \{ U \in \mathrm{U}(\mathscr{E}) \colon U \mathscr{E}_+ \subseteq \mathscr{E}_+ \}$$

is a subsemigroup of $\mathrm{U}(\mathscr{E})$, *and* $S(\mathscr{E}_+, \theta) := S(\mathscr{E}_+) \cap S(\mathscr{E}_+)^\sharp$ *is $\sharp$-invariant. The OS transform defines a $*$-representation* $(\Gamma, \widehat{\mathscr{E}})$ *of the involutive semigroup* $(S(\mathscr{E}_+, \theta), \sharp)$ *by contractions on* $\widehat{\mathscr{E}}$ *which is continuous with respect to the strong operator topologies on* $S(\mathscr{E}_+, \theta)$ *and* $B(\widehat{\mathscr{E}})$.

Proof Clearly, $S := S(\mathscr{E}_+, \theta)$ is $\sharp$-invariant and hence an involutive semigroup. For $\xi, \eta \in \mathscr{E}_+$ we have

$$\langle U\xi, \eta \rangle_\theta = \langle U\xi, \theta\eta \rangle = \langle \xi, U^{-1}\theta\eta \rangle = \langle \xi, \theta U^\sharp \eta \rangle = \langle \xi, U^\sharp \eta \rangle_\theta, \qquad (3.3)$$

and this implies

$$\langle U\xi, U\xi \rangle_\theta = \langle \xi, U^\sharp U\xi \rangle_\theta. \qquad (3.4)$$

Lemma 3.1.2 shows that any $U \in S$ induces a linear operator $\widehat{U}$ on the dense subspace $q(\mathscr{E}_+) \subseteq \widehat{\mathscr{E}}$. Since $U^\sharp U$ is also contained in S, we obtain from Lemma 3.1.2 that $\|\widehat{U^\sharp U}\| \leq \|U^\sharp U\| = 1$. With (3.4) we thus get $\|\widehat{U}\| \leq 1$ so that $\widehat{U}$ extends to a contraction, also denoted $\widehat{U} = \Gamma(U)$, on $\widehat{\mathscr{E}}$. The relation $\widehat{UV} = \widehat{U}\widehat{V}$ for $U, V \in S$ follows on the dense subspace $q(\mathscr{E}_+)$ immediately from the definition, and $\widehat{U}^* = \widehat{U^\sharp}$ is a consequence of (3.3).

The continuity of Γ with respect to the weak operator topology on $B(\widehat{\mathscr{E}})$ follows from the fact that, for $\xi \in \mathscr{E}_+$, the function $\Gamma^{\widehat{\eta}, \widehat{\xi}}(U) := \langle \widehat{\eta}, \widehat{U}\widehat{\xi} \rangle = \langle \theta\eta, U\xi \rangle$ is continuous on S, endowed with the strong operator topology (which equals the weak operator topology). Now [Nel64, Cor. IV.1.18] implies that Γ is strongly continuous. $\qquad \square$

The preceding lemma implies that, for symmetric subsemigroups, reflection positive representations lead by the OS transform to $*$-representations of S by contractions on $\widehat{\mathscr{E}}$:

Proposition 3.3.3 (OS transform of a representation) *If $(U, \mathscr{E})$ is a unitary representation of G_τ on the reflection positive Hilbert space $(\mathscr{E}, \mathscr{E}_+, \theta)$ which is reflection positive with respect to the symmetric subsemigroup $S \subseteq G$, then the OS transform defines a strongly continuous $*$-representation $(\widehat{U}, \widehat{\mathscr{E}})$ of the involutive semigroup $(S, \sharp)$ by contractions.*

Proof The invariance of S under $\sharp$ and the relation $U_{s^\sharp} = U_{\tau(s^{-1})} = \theta U_s^{-1} \theta = U_s^\sharp$ imply that $U_S \subseteq S(\mathscr{E}_+, \theta)$. The remaining assertions now follow immediately from Lemma 3.3.2. $\blacksquare$

Definition 3.3.4 In the context of Proposition 3.3.3, we call $(U, \mathscr{E}, \mathscr{E}_+, \theta)$ a *euclidean realization* of the contractive $*$-representation $(\widehat{U}, \widehat{\mathscr{E}})$ of S.

In Chap. 6 we shall encounter methods to derive by analytic continuation from a $*$-representation $\widehat{U}$ of S (if S has interior points) a unitary representation of the simply connected c-dual group G^c (cf. Example 6.4.2). In this context, we also speak of euclidean realizations of unitary representations of G^c.

Example 3.3.5 For $(G, S, \tau) = (\mathbb{R}, \mathbb{R}_{\geq 0}, -\operatorname{id}_\mathbb{R})$, the situation is particularly simple. Then $U: \mathbb{R}_\tau \to \mathrm{U}(\mathscr{E})$ is a unitary representation and $\widehat{U}: \mathbb{R}_{\geq 0} \to B(\widehat{\mathscr{E}})$ is a continuous one-parameter semigroup of hermitian contractions, hence of the form $\widehat{U}_t = e^{-tH}$ for some selfadjoint positive operator $H = H^* \geq 0$. Then $U_t^c := e^{itH}$ defines a unitary representation U^c of the c-dual group $G^c \cong \mathbb{R}$ on $\widehat{\mathscr{E}}$ related to $\widehat{U}$ by analytic continuation. We shall analyze such examples more closely in Chap. 4.

There is also the following rather weak notion of a reflection positive representation:

Definition 3.3.6 Let (G, H) be a symmetric Lie group and $(\mathscr{E}, \mathscr{E}_+, \theta)$ be a reflection positive Hilbert space. A unitary representation $(U, \mathscr{E})$ of G_τ is called *infinitesimally reflection positive* if

(a) $U_h \mathscr{E}_+ = \mathscr{E}_+$ for every $h \in H$, and
(b) there exists a subspace $\mathscr{D} \subseteq \mathscr{E}^\infty \cap \mathscr{E}_+$ such that $\widehat{\mathscr{D}}$ is dense in $\widehat{\mathscr{E}}$ and $\mathrm{d}U(\mathfrak{q})\mathscr{D} \subset \mathscr{D}$.

Remark 3.3.7 Condition (a) in Definition 3.3.6 implies the existence of a unitary representation $\widehat{U}$ of H on $\widehat{\mathscr{E}}$ given by $\widehat{U}_h = \widehat{U_h}$ (Proposition 3.1.4). Condition (b) ensures that each operator $\mathrm{d}U(x)$, $x \in \mathfrak{q}$, has an OS transform $\widehat{\mathrm{d}U(x)}: \widehat{\mathscr{D}} \to \widehat{\mathscr{D}}$, and one easily verifies the relation $\widehat{\mathrm{d}U}(\mathrm{Ad}(h)x) = \widehat{U}_h \widehat{\mathrm{d}U(x)} \widehat{U}_{h^{-1}}$ for $h \in H$ and $x \in \mathfrak{q}$.

Example 3.3.8 Recall the setting of Theorem 2.5.1, where M is a Riemannian manifold and we obtain the reflection positive Hilbert space $(\mathscr{E}, \mathscr{E}_+, \theta_C)$ with $\mathscr{E} = \mathscr{H}_D$. Since the operator C commutes with the unitary representation of the Lie group

$G := \mathrm{Isom}(M)$ on $L^2(M)$, we obtain a unitary representation $(U^C, \mathscr{E})$ of this group which also contains θ. For the identity component $H := G_0^\theta$, we then have $U_h \mathscr{E}_+ = \mathscr{E}_+$ because elements in H cannot map M_+ to M_-. Now the image $\mathscr{D}$ of $C_c^\infty(M_+)$ in $\mathscr{E}_+$ is a U_H-invariant dense subspace invariant under the action of the Lie algebra $\mathfrak{g}$ of G which acts by Lie derivatives

$$(\mathscr{L}_X f)(m) = \frac{d}{dt}\Big|_{t=0} f(\exp(tX).m),$$

where we identify $\mathfrak{g}$ with a Lie algebra of vector fields on M. We conclude that all requirements of Definition 3.3.6 are satisfied.

Remark 3.3.9 (*Infinitesimally unitary representations*) On the infinitesimal level, the core idea of reflection positivity is easily seen. Starting with a symmetric Lie algebra $(\mathfrak{g}, \tau)$, we obtain the corresponding decomposition $\mathfrak{g} = \mathfrak{h} \oplus \mathfrak{q}$ and form the dual Lie algebra $\mathfrak{g}^c := \mathfrak{h} \oplus i\mathfrak{q} \subseteq \mathfrak{g}_\mathbb{C}$.

Let $(\mathscr{D}, \mathscr{D}_+, \theta)$ be a complex *reflection positive pre-Hilbert space* (defined as in Definition 2.1.1 but omitting the completeness of $\mathscr{E}$ and the closedness of $\mathscr{E}_+$) and π be a representation of $\mathfrak{g}$ on $\mathscr{D}$ by skew-symmetric operators. We also assume that $\theta \pi(x) \theta = \pi(\tau x)$ for $x \in \mathfrak{g}$ and that $\mathscr{D}_+$ is $\mathfrak{g}$-invariant. Then complex linear extension leads to a representation of $\mathfrak{g}^c$ on $\mathscr{D}_+$ by operators which are skew-symmetric with respect to the twisted scalar product $\langle \cdot, \cdot \rangle_\theta$. By the OS transform, we then obtain an infinitesimally unitary representation of $\mathfrak{g}^c$ on the associated pre-Hilbert space $\widehat{\mathscr{D}}$ via

$$\pi^c(x + iy) := \widehat{\pi(x)} + i\widehat{\pi(y)}.$$

This is the basic idea behind the reflection positivity correspondence between infinitesimally unitary representations of $\mathfrak{g}$ on $\mathscr{E}$ and $\mathfrak{g}^c$ on $\widehat{\mathscr{E}}$.

What this simple picture completely ignores are issues of integrability and essential selfadjointness of operators. There are various natural ways to address these problems. Important first steps in this direction have been undertaken by Klein and Landau in [KL81, KL82], and Fröhlich, Osterwalder and Seiler introduced in [FOS83] the concept of a virtual representation, which was developed in greater generality by Jorgensen in [Jo86, Jo87]. We shall return to these issues in Chap. 7.

3.4 Reflection Positive Functions

Definition 3.4.1 Let V be a real vector space and (G, τ) be a symmetric Lie group. We recall the group $G_\tau = G \rtimes \{\mathrm{id}_G, \tau\}$ from Definition 3.2.1.

(a) A function $\varphi : G_\tau \to \mathrm{Bil}(V)$ (the space of bilinear forms on V) is called *reflection positive with respect to the subset* $G_+ \subseteq G$ if

(RP1) φ is positive definite (cf. Sect. A.1) and

(RP2) the kernel $(s, t) \mapsto \varphi(st^\sharp \tau) = \varphi(s\tau t^{-1})$ is positive definite on G_+.

(b) A τ-invariant function $\varphi : G \to \mathrm{Bil}(V)$ is called *reflection positive with respect to* G_+ if the extension $\widehat{\varphi}$ of φ to G_τ by $\widehat{\varphi}(g, \tau) := \varphi(g)$ has this property, i.e., if the kernel $(\varphi(st^\sharp))_{s,t \in G_+}$ is positive definite.

(c) A function $\varphi : G \to \mathrm{Bil}(V)$ is called *reflection positive with respect to the symmetric subsemigroup* $S \subseteq G$ if

(RP1) φ is positive definite and τ-invariant and
(RP2)   the kernel $(\varphi(st^\sharp))_{s,t \in S}$ is positive definite on S, i.e., the restriction $\varphi|_S$ is a positive definite function on the involutive semigroup $(S, \sharp)$.

This can also be phrased as the requirement that the kernel $K(g, h) := \varphi(gh^{-1})$ on G is reflection positive with respect to the symmetric subsemigroup $S \subseteq (G, \tau)$ in the sense of Definition 2.4.1.

Remark 3.4.2 Let $\varphi : G_\tau \to \mathrm{Bil}(V)$ be a positive definite function, so that the kernel $K((x, v), (y, w)) := \varphi(xy^{-1})(v, w)$ on $G_\tau \times V$ is positive definite. The involution τ acts on $G_\tau \times V$ by $\tau.(g, v) := (g\tau, v)$ and the corresponding kernel $K^\tau((x, v), (y, w)) := K((x, v), (y\tau, w)) = \varphi(x\tau y^{-1})(v, w)$ is positive definite on $G_+ \times V$ if and only if φ is reflection positive in the sense of Definition 2.4.1.

From Lemma 2.4.2(c) it follows that the corresponding space $\widehat{\mathscr{E}}$ can be identified with $\mathscr{H}_{K^\tau} \subseteq (V^*)^{G_+}$ such that

$$q : \mathscr{E}_+ \to \mathscr{H}_{K^\tau}, \quad q(f)(g) := f(\tau(g)), \quad g \in G_+.$$

The following lemma shows that positive definite functions on G extend canonically to G_τ if they are τ-invariant:

Lemma 3.4.3 *Let V be a real vector space and let (G, τ) be a symmetric Lie group. Then the following assertions hold:*

(i) *If $\varphi : G \to \mathrm{Bil}(V)$ is a positive definite function which is τ-invariant in the sense that $\varphi \circ \tau = \varphi$, then $\widehat{\varphi}(g, \tau) := \varphi(g)$ defines an extension to G_τ which is positive definite and τ-biinvariant.*

(ii) *Let $(U, \mathscr{H})$ be a unitary representation of G_τ, let $\theta := U_\tau$, let $j : V \to \mathscr{H}$ be a linear map, and let $\varphi(g)(v, w) = \langle j(v), U_g j(w) \rangle$ be the corresponding $\mathrm{Bil}(V)$-valued positive definite function. Then the following are equivalent:*

  (a) $\theta j(v) = j(v)$ *for every* $v \in V$.
  (b) φ *is τ-biinvariant.*
  (c) φ *is left τ-invariant.*

Proof (i) From the GNS construction (Proposition A.1.6), we obtain a continuous unitary representation $(U, \mathscr{H})$ of G and a linear map $j : V \to \mathscr{H}$ such that

$$\varphi(g)(v, w) = \langle j(v), U_g j(w) \rangle \quad \text{for} \quad g \in G, v, w \in V.$$

As $\varphi(g)(v, w) = \varphi(\tau(g))(v, w)$, the uniqueness in the GNS construction provides a unitary operator $\theta : \mathscr{H} \to \mathscr{H}$ with

$$\theta U_g j(v) = U_{\tau(g)} j(v) \quad \text{for} \quad g \in G, v \in V.$$

Note that θ fixes each $j(v)$. Therefore $U_\tau := \theta$ defines an extension of U to a unitary representation of G_τ on $\mathcal{H}$. Hence $\psi(g)(v, w) = \langle j(v), U_g j(w) \rangle$ defines a positive definite τ-biinvariant Bil(V)-valued function on G_τ with $\psi|_G = \varphi$.

(ii) Clearly, (a) $\Rightarrow$ (b) $\Rightarrow$ (c). It remains to show that (c) implies (a). So we assume that $\varphi(\tau g) = \varphi(g)$ for $g \in G_\tau$. This means that, for every $v, w \in V$, we have

$$\langle j(v), U_g j(w) \rangle = \varphi(g)(v, w) = \varphi(\tau g)(v, w) = \langle j(v), \theta U_g j(w) \rangle = \langle \theta j(v), U_g j(w) \rangle.$$

Since $U_{G_\tau} j(V)$ is total in $\mathcal{H}$, this implies that $\theta j(v) = j(v)$ for every $v \in V$. $\square$

Remark 3.4.4 If $S \subseteq G$ is a symmetric subsemigroup, then a function $\varphi: G \to$ Bil(V) is reflection positive with respect to S if and only if its τ-biinvariant extension to G_τ (Lemma 3.4.3) is reflection positive with respect to $G_+ = S$.

Theorem 3.4.5 (GNS construction for reflection positive functions) *Let V be a real vector space, let $(U, \mathcal{E})$ be a unitary representation of G_τ and put $\theta := U_\tau$. Then the following assertions hold:*

(i) *If $(U, \mathcal{H}, j, V)$ is reflection positive with respect to G_+, then*

$$\varphi(g)(v, w) := \langle j(v), U_g j(w) \rangle, \quad g \in G_\tau, v, w \in V,$$

is a reflection positive Bil(V)-valued function.

(ii) *If $\varphi: G_\tau \to$ Bil(V) is a reflection positive function with respect to G_+, then the corresponding GNS representation $(U^\varphi, \mathcal{H}_\varphi, j, V)$ is a reflection positive representation, where $\mathcal{E} := \mathcal{H}_\varphi \subseteq \mathbb{C}^{G_\tau \times V}$ is the Hilbert subspace with reproducing kernel $K((x, v), (y, w)) := \varphi(xy^{-1})(v, w)$ on which G_τ acts by*

$$(U_g^\varphi f)(x, v) := f(xg, v).$$

Further, $\mathcal{E}_+ := [\![U_{G_+}^{-1} j(V)]\!]$ and $\hat{\mathcal{E}} \cong \mathcal{H}_{K^\tau}$ for the kernel $K^\tau(s, t) := \varphi(s\tau t^{-1})$ on G_+, where $q: \mathcal{E}_+ \to \mathcal{H}_{K^\tau}, q(f)(g) := f(g\tau)$.

Proof (i) For $s, t \in G_+$, we have

$$\varphi(s\tau t^{-1})(v, w) = \langle j(v), U_{s\tau t^{-1}} j(w) \rangle = \langle U_{s^{-1}} j(v), U_\tau U_{t^{-1}} j(w) \rangle$$
$$= \langle \theta U_{s^{-1}} j(v), U_{t^{-1}} j(w) \rangle,$$

so that the kernel $(\varphi(s\tau t^{-1}))_{s,t \in G_+}$ is positive definite by Proposition A.1.6.

(ii) Recall the relation $\varphi(g)(v, w) = \langle j(v), U_g j(w) \rangle$ for $g \in G, v, w \in V$ from Proposition A.1.6. Moreover, $(\theta f)(x, v) = f(x\tau, v)$, and

$$\langle \theta U_{s^{-1}}^\varphi j(v), U_{t^{-1}}^\varphi j(w) \rangle = \langle j(v), U_{s\tau t^{-1}}^\varphi j(w) \rangle = \varphi(s\tau t^{-1})(v, w),$$

so that the positive definiteness of the kernel $(\varphi(stt^{-1}))_{s,t \in G_+}$ implies that we obtain with $\mathcal{E} = \mathcal{H}_\varphi$ and $\mathcal{E}_+ := [\![(U_{G_+}^\varphi)^{-1} j(V)]\!]$ a reflection positive Hilbert space $(\mathcal{E}, \mathcal{E}_+, \theta)$. The remaining assertions follow from Remark 3.4.2. $\square$

Definition 3.4.6 Let $S \subseteq (G, \tau)$ be a symmetric subsemigroup. A triple $(U, \mathcal{E}, \mathcal{V})$, where $(U, \mathcal{E})$ is a unitary representation of G_τ and $\mathcal{V} \subseteq \mathcal{E}$ is a G-cyclic subspace fixed pointwise by $\theta = U_\tau$ is said to be a *a reflection positive $\mathcal{V}$-cyclic representation* if the closed subspace $\mathcal{E}_+ := [\![U_S \mathcal{V}]\!]$ is θ-positive. If, in addition, $\mathcal{V} = \mathbb{C}\xi$ is one-dimensional, then we call the triple $(\pi, \mathcal{E}, \xi)$ *a reflection positive cyclic representation*.

Corollary 3.4.7 (Reflection positive GNS construction—operator-valued case) *Let S be a symmetric subsemigroup of (G, τ).*

(i) *If $(U, \mathcal{E}, \mathcal{V})$ is an $\mathcal{V}$-cyclic reflection positive representation of G_τ and $P \colon \mathcal{E} \to \mathcal{V}$ the orthogonal projection, then $\varphi(g) := P U_g P^*$ is a reflection positive function on G with $\varphi(e) = \mathbf{1}_{\mathcal{V}}$.*

(ii) *Let $\varphi \colon G \to B(\mathcal{V})$ is a reflection positive function with respect to S on G with $\varphi(e) = \mathbf{1}_{\mathcal{V}}$ and let $\mathcal{H}_\varphi \subseteq \mathcal{V}^G$ be the Hilbert subspace with reproducing kernel $K(x, y) := \varphi(xy^{-1})$ on which G acts by $(U^\varphi(g)f)(x) := f(xg)$ and τ by $\theta f := f \circ \tau$. We identify $\mathcal{V}$ with the subspace $\mathrm{ev}_e^* \mathcal{V} \subseteq \mathcal{H}_\varphi$. Then $(U^\varphi, \mathcal{H}_\varphi, \mathcal{V})$ is a $\mathcal{V}$-cyclic reflection positive representation and we have an S-equivariant unitary map*

$$\Gamma \colon \widehat{\mathcal{E}} \to \mathcal{H}_{\varphi|_S}, \quad \Gamma(\widehat{f}) = f|_S \quad for \quad f \in \mathcal{E}_+ = [\![U_S^\varphi \mathcal{V}]\!].$$

Proof (i) To match this with Theorem 3.4.5(i), we put $V := \mathcal{V}$ and consider the inclusion map $j \colon V \to \mathcal{H}$. Then $\varphi(g) \in \mathrm{Bil}(V)$ corresponds to the operator $j^* U_g j = P U_g P^* \in B(V)$. Therefore φ is positive definite with $\varphi(e) = \mathbf{1}$. That φ is τ-invariant follows from $\theta|_{\mathcal{V}} = \mathrm{id}_{\mathcal{V}}$ (cf. Lemma 3.4.3).

(ii) We use the second half of Example A.1.8, i.e., the special case of Proposition A.1.6 dealing with operator-valued positive definite functions, and identify $\mathcal{V}$ with $\mathrm{ev}_e^* \mathcal{V} \subseteq \mathcal{H}_\varphi$. Lemma 3.4.3 implies that θ fixes $\mathcal{V}$ pointwise.

To see that $\mathcal{E}_+ := [\![U_S^\varphi \mathcal{V}]\!]$ is θ-positive, we note that

$$\theta(\mathcal{E}_+) = [\![U_{\tau(S)}^\varphi \theta \mathcal{V}]\!] = [\![U_{S^{-1}}^\varphi \mathcal{V}]\!],$$

and this subspace is θ-positive by Theorem 3.4.5(ii). Therefore $\mathcal{E}_+$ is also θ-positive (Remark 2.2.2). From Theorem 3.4.5(ii) we further derive that

$$\theta(\mathcal{E}_+) \to \mathcal{H}_{K^\tau} \subseteq (V^*)^S, \quad f \mapsto (f \circ \tau)|_S = \theta(f)|_S$$

induces a unitary isomorphism $\widehat{\mathcal{E}} \to \mathcal{H}_{K^\tau}, \widehat{f} \mapsto f|_S$, and this implies that Γ is unitary. $\square$

Corollary 3.4.8 *Let $S \subseteq (G, \tau)$ be a symmetric subsemigroup.*

(i) *If $(U, \mathscr{E}, \xi)$ is a cyclic reflection positive representation of G_τ, then $U^\xi(g) :=$ $\langle \xi, U_g \xi \rangle$ is a reflection positive function on G.*
(ii) *If φ is a reflection positive function on G, then $(U^\varphi, \mathscr{H}_\varphi, \varphi)$ is a cyclic reflection positive representation.*

The following result characterizes reflection positive representations for which $(\mathscr{E}, \mathscr{E}_+, \theta)$ is of Markov type.

Proposition 3.4.9 *Let $(U, \mathscr{E})$ be a reflection positive unitary representation on $(\mathscr{E}, \mathscr{E}_+, \theta)$ with respect to the unital symmetric subsemigroup $S \subseteq (G, \tau)$. Let $P_0: \mathscr{E} \to \mathscr{E}_0$ be the orthogonal projection and consider the reflection positive definite function $\varphi(g) = P_0 U_g P_0$. Then the following assertions hold:*

(a) *If $\varphi|_S$ is multiplicative and $\mathscr{E}_+ = [\![U_S \mathscr{E}_0]\!]$, then $(\mathscr{E}, \mathscr{E}_+, \theta)$ is of Markov type.*
(b) *If $(\mathscr{E}, \mathscr{E}_+, \theta)$ is of Markov type and $\Gamma := q|_{\mathscr{E}_0}: \mathscr{E}_0 \to \widehat{\mathscr{E}}$ is the corresponding unitary isomorphism, then $\varphi|_S$ is multiplicative and $\varphi(s) = \Gamma^* \widehat{U}_s \Gamma$ for $s \in S$, i.e., Γ intertwines $\varphi|_S$ with the contraction representation $(\widehat{U}, \widehat{\mathscr{E}})$ of S.*

Proof That φ is reflection positive follows from Corollary 3.4.7(i).

(a) By Corollary 3.4.7(ii), the restriction map $\Gamma: \mathscr{E}_+ \to \mathscr{H}_{\varphi|_S}, \widehat{f} \mapsto f|_S$ is a unitary S-intertwining operator. From $\mathscr{E}_+ = [\![U_S \mathscr{E}_0]\!]$ it follows that $\widehat{\mathscr{E}} = [\![\widehat{U}_S q(\mathscr{E}_0)]\!]$, so that the multiplicativity of $\varphi|_S$ implies that $\Gamma(\mathscr{E}_0) = \widehat{\mathscr{E}}$ (Lemma A.1.9), i.e., $(\mathscr{E}, \mathscr{E}_+, \theta)$ is of Markov type.

(b) Let $\mathscr{K} \subseteq \mathscr{H}$ be the U-invariant closed subspace generated by $\mathscr{E}_0$ and let $(\mathscr{E}_0)^G$ denote the linear space of all maps $G \to \mathscr{E}_0$. Then the map

$$\Phi: \mathscr{K} \to (\mathscr{E}_0)^G, \quad \Phi(\xi)(g) := P_0 U_g \xi$$

is an equivalence of the representation U of G on $\mathscr{K}$ with the GNS representation defined by φ (Proposition A.1.6). Further, the representation $\widehat{U}$ of S on $\widehat{\mathscr{E}}$ is equivalent to the GNS representation defined by $\varphi|_S$, where the map $q: \mathscr{E}_+ \to \widehat{\mathscr{E}}$ corresponds to the restriction $f \mapsto f|_S$ (Corollary 3.4.7(ii)). The inclusion $\iota: \mathscr{E}_0 \hookrightarrow \mathscr{H}_\varphi$ is given by $\iota(\xi)(g) = P_0 U_g \xi = \varphi(g)\xi$ for $g \in G$, and likewise the inclusion $\widehat{\iota}: \mathscr{E}_0 \hookrightarrow \mathscr{H}_{\varphi|_S}$ is given by $\widehat{\iota}(\xi) = \varphi \cdot \xi$. Lemma 2.3.2 implies the surjectivity of $\widehat{\iota}$. In view of Lemma A.1.9, this is equivalent to the multiplicativity of $\varphi|_S$.

Recall $q = \Gamma \circ P_0|_{\mathscr{E}_+}$ from Lemma 2.3.2. For $s \in S$, the relation $\widehat{U}_s \circ q = q \circ U_s|_{\mathscr{E}_+}$ leads to $\widehat{U}_s \Gamma P_0|_{\mathscr{E}_+} = \Gamma P_0 U_s|_{\mathscr{E}_+}$, so that $\Gamma^* \widehat{U}_s \Gamma = P_0 U_s P_0 = \varphi(s)$, i.e., Γ intertwines $\varphi(s)$ with $\widehat{U}_s$.

Example 3.4.10 Let (G, τ) be a symmetric Lie group and let $\sigma: G_\tau \to \mathrm{Diff}(M)$ be a smooth right action of G_τ on the manifold M. Then $\tau_M := \sigma_\tau$ is an involutive diffeomorphism of M. Further, let $K: M \times M \to B(V)$ be a G-invariant reflection positive kernel with respect to (M, M_+, τ_M) (Definition 2.4.1), where $M_+ \subseteq M$ is a H-invariant subset.

Then $U_g f := f \circ \sigma_g$ defines a unitary representation of G on $\mathcal{H}_K$. It clearly satisfies

$$K_x \circ U_g = K_{x.g} \quad \text{and thus} \quad U_g K_x^* = K_{x.g^{-1}}^*. \tag{3.5}$$

Here the unitarity of U follows from the G-invariance of K, and the H-invariance of $\mathscr{E}_+$ follows from the H-invariance of M_+ and (3.5).

A special case of this construction arises for $G = M$ and right-invariant kernels of the form $K(x, y) = \varphi(xy^{-1})$, where $\varphi : G \to B(V)$ is a reflection positive function. Here the τ-invariance of K is equivalent to the relation $\varphi \circ \tau = \varphi$.

Notes

Section 3.1: Variants of Lemma 3.1.2 also appear in [JÓl00].

Section 3.3: In the context of intervals in the real line which we discuss in Chap. 4, the notion of reflection positive functions already appears in [KL81], where such functions are called (OS)-positive.

Proposition 3.3.3 is already in [JÓl00].

A version of Proposition 3.4.9 for the case $(\mathbb{R}, \mathbb{R}_+, -\mathrm{id}_\mathbb{R})$ can already be found in [Kl77] (see also [JT17, Sect. 7]).

Reflection positivity for the lattice $G = \mathbb{Z}^d$ and $\tau(x_0, \mathbf{x}) = (-x_0, \mathbf{x})$ is discussed in the context of correlation functions by Usui in [Us12].

Chapter 4
Reflection Positivity on the Real Line

After providing the conceptual framework for reflection positive representations in the preceding two chapters, we now turn to the fine points of reflection positivity on the additive group $(\mathbb{R}, +)$. Although this Lie group is quite trivial, reflection positivity on the real line has many interesting facets and is therefore quite rich. We thus describe its main features in this and the subsequent chapter. As reflection positive functions play a crucial role, we start in Sect. 4.1 with reflection positive functions on intervals $(-a, a) \subseteq \mathbb{R}$. Here we already encounter the main feature of reflection positivity dealing with two different notions of positivity, one related to the group structure on $\mathbb{R}$ and the other related to the $*$-semigroup structure on $\mathbb{R}_+$, resp., the convex structure of intervals. All this is linked to representation theory in Sect. 4.2, where we start our investigation of reflection positive representations of the symmetric semigroup $(\mathbb{R}, \mathbb{R}_+, - \mathrm{id}_\mathbb{R})$. These are unitary one-parameter groups $(U_t)_{t \in \mathbb{R}}$ on a reflection positive Hilbert space $(\mathscr{E}, \mathscr{E}_+, \theta)$ satisfying $U_t \mathscr{E}_+ \subseteq \mathscr{E}_+$ for $t > 0$ and $\theta U_t \theta = U_{-t}$ for $t \in \mathbb{R}$. On $\widehat{\mathscr{E}}$ this leads to a semigroup $(\widehat{U}_t)_{t \geq 0}$ of hermitian contractions. The main result in Sect. 4.2 is that the OS transform "commutes with reduction", where reduction refers to the passage to the fixed points of U and $\widehat{U}$ in $\mathscr{E}$ and $\widehat{\mathscr{E}}$, respectively (Proposition 4.2.6). Reflection positive functions for $(\mathbb{R}, \mathbb{R}_+, - \mathrm{id}_\mathbb{R})$ are classified in terms of integral representations in Sect. 4.3. We shall see in particular that any hermitian contraction semigroup $(C_t)_{t \geq 0}$ on a Hilbert space $\mathscr{H}$ has a so-called minimal dilation represented by the reflection positive function $\psi(t) := C_{|t|}$. We also provide a concrete model for this dilation on the space $\mathscr{E} = L^2(\mathbb{R}, \mathscr{H})$ with $(U_t f)(p) = e^{itp} f(p)$, where $\mathscr{E}_+ = L^2_+(\mathbb{R}, \mathscr{H})$ is the positive spectral subspace for the translation group, which is, by the Laplace transform, isomorphic to the $\mathscr{H}$-valued Hardy space $H^2(\mathbb{C}_+, \mathscr{H})$ on the right half plane $\mathbb{C}_+ = \mathbb{R}_+ + i\mathbb{R}$. We conclude this chapter by showing that, for any reflection positive one-parameter group for which $\mathscr{E}_+$ is cyclic and fixed points are trivial, the space $\mathscr{E}_+$ is outgoing in the sense of Lax–Phillips scattering theory (Proposition 4.4.2). This establishes a remarkable connection between reflection positivity and scattering theory that leads to a normal form of reflection positive one-parameter groups by

© The Author(s) 2018   35
K.-H. Neeb and G. Ólafsson, *Reflection Positivity*, SpringerBriefs in Mathematical Physics, https://doi.org/10.1007/978-3-319-94755-6_4

translations on spaces of the form $\mathscr{E} = L^2(\mathbb{R}, \mathscr{H})$ with $\mathscr{E}_+ = L^2(\mathbb{R}_+, \mathscr{H})$. Applying the Fourier transform to our concrete dilation model leads precisely to this normal form.

4.1 Reflection Positive Functions on Intervals

Before we turn to representation theoretic issues, we briefly discuss reflection positive functions on open intervals in $\mathbb{R}$. There are two natural types of positive definiteness conditions for functions on real intervals. The first one comes from the additive group $(\mathbb{R}, +)$, for which a function $\varphi \colon \mathbb{R} \to \mathbb{C}$ is positive definite if and only if the kernel $(\varphi(x - y))_{x,y \in \mathbb{R}}$ is positive definite. This condition makes also sense on symmetric intervals of the form $(-a/2, a/2)$ if f is defined on $(-a, a)$. Bochner's Theorem (Theorem A.2.1) asserts that a continuous function on the additive group $\mathbb{R}$ is positive definite if and only if it is the Fourier transform $\varphi(x) = \int_{\mathbb{R}} e^{-ix\lambda} \, d\mu(\lambda)$ of a bounded positive Borel measure μ on $\mathbb{R}$.

The second type makes sense for functions $\varphi \colon (a, b) \to \mathbb{C}$ on any real interval and requires that the kernel $\left(\varphi\left(\frac{x+y}{2}\right)\right)_{a<x,y<b}$ is positive definite. Widder's Theorem below asserts that this is equivalent to f being a Laplace transform of a positive Borel measure μ on $\mathbb{R}$. For $(a, b) = (0, \infty)$, this is precisely the condition of positive definiteness on the $*$-semigroup $(0, \infty)$ with the trivial involution $t^* = t$ for $t > 0$.

Theorem 4.1.1 (Widder; [Wi34], [Wi46, Theorem VI.21]) *Let* $-\infty \leq a < b \leq \infty$. *A function* $\varphi \colon (a, b) \to \mathbb{R}$ *is positive definite in the sense that the kernel* $\varphi\left(\frac{x+y}{2}\right)$ *is positive definite if and only if there exists a positive Borel measure* μ *on* $\mathbb{R}$ *such that*

$$\varphi(t) = \mathscr{L}(\mu)(t) := \int_{\mathbb{R}} e^{-\lambda t} \, d\mu(\lambda) \quad for \quad t \in (a, b).$$

This implies in particular that φ *is analytic.*

The following theorem provides a characterization of functions $\psi \colon (0, \infty) \to \mathbb{R}$ which are *completely monotone*, i.e., $(-1)^k \psi^{(k)} \geq 0$ for $k = 1, 2, 3, \ldots$ (see [JNO18, Theorem 3.6], [SSV10, Theorem 1.4], [Wi46, Theorem IV.12b]). Its most remarkable point is that it characterizes the global property of positive definiteness on the additive semigroup $(0, \infty)$ in terms of the infinitesimal condition of being completely monotone.

Theorem 4.1.2 (Hausdorff–Bernstein–Widder) *For a function* $\varphi \colon (0, \infty) \to [0, \infty)$, *the following are equivalent:*

(i) φ *is completely monotone.*
(ii) φ *is a Laplace transform of a positive Borel measure on* $[0, \infty)$.
(iii) φ *is decreasing and positive definite on the* $*$-semigroup $((0, \infty), \mathrm{id})$.

After these preparations, we now turn to reflection positive functions on intervals.

Definition 4.1.3 Let $a \in (0, \infty]$ and consider the interval $(-a, a)$, endowed with the reflection $\tau(t) = -t$ about the midpoint. We call a function $\varphi \colon (-a, a) \to \mathbb{R}$ *reflection positive* if both kernels

$$\varphi\left(\frac{t-s}{2}\right)_{-a<s,t<a} \quad \text{and} \quad \varphi\left(\frac{t+s}{2}\right)_{0<s,t<a} \tag{4.1}$$

are positive definite. For the kernel $K(s, t) := \varphi\left(\frac{t-s}{2}\right)$, this corresponds to the situation of Definition 2.4.1 with $X = (-a, a)$, $X_+ = (0, a)$ and $\tau(x) = -x$.

Since the kernels in (4.1) are both hermitian, reflection positive functions on $(-a, a)$ satisfy $\varphi(-t) = \overline{\varphi(t)} = \varphi(t)$. Therefore Widder's Theorem 4.1.1 provides a positive Borel measure μ on $\mathbb{R}$ with

$$\varphi(t) = \mathscr{L}(\mu)(|t|) = \int_{\mathbb{R}} e^{-\lambda|t|} \, d\mu(\lambda) \quad \text{for} \quad |t| < a. \tag{4.2}$$

Conversely, for such functions the kernel $\varphi\left(\frac{t+s}{2}\right)$ is positive definite on $(0, a)$. Therefore φ is reflection positive if and only if the kernel $\varphi\left(\frac{t-s}{2}\right)$ is positive definite on $(-a, a)$. So the main point is to relate this condition to properties of the measure μ.

Example 4.1.4 (a) For $\lambda \geq 0$, the functions $\varphi_\lambda(t) := e^{-\lambda|t|}$ are positive definite (Example 2.4.3(a)). Therefore $\mathscr{L}(\mu)(|t|)$ is reflection positive if μ is supported by $[0, \infty)$.

(b) Basic examples of positive definite β-periodic functions on $\mathbb{R}$ are given by

$$f_\lambda(t) = e^{-t\lambda} + e^{-(\beta-t)\lambda} = 2e^{-\beta\lambda/2} \cosh((\tfrac{\beta}{2} - t)\lambda) \quad \text{for} \quad 0 \leq t \leq \beta, \lambda \geq 0$$

(Example 2.4.3(b)). For $|t| < \beta$, we then have

$$f_\lambda(t) = f_\lambda(|t|) = e^{-|t|\lambda} + e^{-(\beta-|t|)\lambda} = e^{-|t|\lambda} + e^{-\beta\lambda}e^{|t|\lambda}. \tag{4.3}$$

Hence, for reflection positivity on a finite interval $(-\beta, \beta)$, it is not necessary that the measure μ in (4.2) is supported by the positive half line, as in (a).

From the positive definiteness of the functions f_λ for every $\beta > 0$, we conclude that, for a fixed $a > 0$ and a positive Borel measure μ on $[0, \infty) \times [a, \infty)$, the function

$$f(t) := \int_{[0,\infty)\times[a,\infty)} e^{-\lambda|t|} + e^{-\beta\lambda}e^{\lambda|t|} \, d\mu(\lambda, \beta) \tag{4.4}$$

is reflection positive on $(-a, a)$ whenever the integrals are finite.

The proof of the following theorem ([JNO18, Theorem 5.8]) uses Pólya's classical result relating positive definiteness of real-valued functions on $\mathbb{R}_+$ with convexity ([Luk70, Theorem 4.3.1]) and provides a sufficient conditions for positive definiteness.

Theorem 4.1.5 (Characterization of reflection positive functions on $[-a, a]$) *Fix $a > 0$ and let μ be a finite positive Borel measure on $\mathbb{R}$ for which $\varphi(t) := \mathscr{L}(\mu)(|t|)$ exists for $|t| \leq a$.*

(i) *If the left-sided derivative in a satisfies $\mathscr{L}(\mu)'(a-) \leq 0$, then φ is reflection positive on $[-a, a]$ and extends to a symmetric positive definite function on $\mathbb{R}$.*

(ii) *If φ is reflection positive on $[-a, a]$ and non-constant, then there exists a number $b \in (0, a]$ with $\mathscr{L}(\mu)'(b-) < 0$.*

Remark 4.1.6 If b is as in (ii), then it follows from part (a) that the positive definite function $\varphi|_{[-b,b]}$ extends to a positive definite function on $\mathbb{R}$. But this extension does not have to coincide with φ if $b < a$.

Remark 4.1.7 (The β-periodic case) We consider the group $G = (\mathbb{R}, +)$ and the open interval $G_+ := (0, \beta/2)$. Then a symmetric continuous function $f: \mathbb{R} \to \mathbb{C}$ is reflection positive with respect to G_+ if it is positive definite and the kernel $\left(f(t + s)\right)_{0 < s,t < \beta/2}$ is positive definite (Definition 3.4.1(b)), which implies that f is reflection positive on the interval $(-\beta, \beta)$ (Definition 4.1.3). As f is symmetric and β-periodic, it is also symmetric with respect to $\beta/2$, i.e., $f(\beta - t) = f(t)$. The latter relation implies that the corresponding measure μ on $\mathbb{R}$ satisfies $d\mu(-\lambda) = e^{-\beta\lambda}d\mu(\lambda)$, hence has the form

$$d\mu(\lambda) = d\mu_+(\lambda) + e^{\beta\lambda}d\mu_+(-\lambda) \tag{4.5}$$

for a measure μ_+ on $\mathbb{R}_{\geq 0}$. We thus obtain the integral representation

$$f(t) = \int_0^\infty e^{-t\lambda} + e^{-(\beta-t)\lambda}\, d\mu_+(\lambda) \quad \text{for} \quad 0 \leq t \leq \beta,$$

which determines f by β-periodicity. That, conversely, all such functions are reflection positive follows from Example 2.4.3(b). Note that $f|_{[0,\beta]}$ is convex and symmetric with respect to $\beta/2$, where it has a global minimum. In particular Theorem 4.1.5 only applies to the restriction of f to the interval $[-\beta/2, \beta/2]$ which also determines f by β-periodicity.

4.2 Reflection Positive One-Parameter Groups

We now turn to reflection positivity on the whole real line $X = \mathbb{R}$ with respect to the right half line $X_+ = \mathbb{R}_{\geq 0} = [0, \infty)$ which is an additive $*$-semigroup with $s^* = s$. Therefore reflection positive functions provide close relations between unitary representations of $\mathbb{R}$ and one-parameter semigroups of hermitian contractions. Specializing Definition 3.3.1 to the symmetric semigroup $(\mathbb{R}, \mathbb{R}_+, -\mathrm{id}_\mathbb{R})$, we obtain:

Definition 4.2.1 A *reflection positive* unitary one-parameter group on the reflection positive Hilbert space $(\mathscr{E}, \mathscr{E}_+, \theta)$ is a strongly continuous unitary one-parameter group $(U_t)_{t \in \mathbb{R}}$ on $\mathscr{E}$ for which $\mathscr{E}_+$ is invariant under U_t for $t > 0$ and $\theta U_t \theta = U_{-t}$ for $t \in \mathbb{R}$.

From Proposition 3.3.3 we immediately obtain:

Proposition 4.2.2 *If* $(U_t)_{t \in \mathbb{R}}$ *is a reflection positive unitary one-parameter group on* $(\mathscr{E}, \mathscr{E}_+, \theta)$, *then* $(\widehat{U}_t)_{t \geq 0}$ *is a strongly continuous one-parameter semigroup of symmetric contractions on* $\widehat{\mathscr{E}}$.

Definition 4.2.3 In the context of Proposition 4.2.2, we call the quadruple $(\mathscr{E}, \mathscr{E}_+, \theta, U)$ a *euclidean realization* of the contraction semigroup $(\widehat{U}, \widehat{\mathscr{E}})$. Writing $\widehat{U}_t = e^{-tH}$ with a $H = H^* \geq 0$, we obtain by analytic continuation the unitary one-parameter group $U_t^c := e^{itH}$ (Example 3.3.5). Accordingly, we also speak of a euclidean realization of $(U_t^c)_{t \in \mathbb{R}}$.

We shall see in Proposition 4.3.5 below that every strongly continuous contraction semigroup on a Hilbert space has a euclidean realization, but there are many non-equivalent ones with different sizes and different specific properties (Example 4.3.8).

The following lemma provides a criterion for the density of a subspace of $\widehat{\mathscr{E}}$. We shall use it to verify that certain operators on $\widehat{\mathscr{E}}$ are densely defined.

Lemma 4.2.4 *Let* $(U_t)_{t \in \mathbb{R}}$ *be a reflection positive unitary one-parameter group on* $(\mathscr{E}, \mathscr{E}_+, \theta)$. *If* $\mathscr{D} \subseteq \mathscr{E}_+$ *is a subspace invariant under the operators* $(U_t)_{t > 0}$, *for which*

$$\mathscr{E}_+^0 := \{v \in \mathscr{E}_+ \colon (\exists T > 0)\ U_T v \in \mathscr{D}\}$$

is dense in $\mathscr{E}_+$, *then* $\widehat{\mathscr{D}} \subseteq \widehat{\mathscr{E}}$ *is dense.*

Proof For $w \in \mathscr{E}_+^0$ there exists a $T > 0$ with $U_T w \in \mathscr{D}$, and this implies that $\widehat{U}_t \widehat{w} \in \widehat{\mathscr{D}}$ for $t \geq T$. Since the curve $\mathbb{R}_+ \to \widehat{\mathscr{E}}, t \mapsto \widehat{U}_t w$, is analytic, $\widehat{U}_t w \in \overline{\widehat{\mathscr{D}}}$ for every $t > 0$, and therefore $w \in \overline{\widehat{\mathscr{D}}}$ follows from the strong continuity of the semigroup $(\widehat{U}_t)_{t \geq 0}$ (Proposition 4.2.2). As $\mathscr{E}_+^0$ is dense in $\mathscr{E}_+$, it follows that $\widehat{\mathscr{D}}$ is dense in $\widehat{\mathscr{E}}$. $\square$

Remark 4.2.5 (Reduction to the $\mathscr{E}_0$-*cyclic case if* $\widehat{\mathscr{E}_0}$ *is cyclic in* $\widehat{\mathscr{E}}$) Assume that $(U_t)_{t \in \mathbb{R}}$ is reflection positive on $(\mathscr{E}, \mathscr{E}_+, \theta)$ and that $q(\mathscr{E}_0)$ is $\widehat{U}$-cyclic in $\widehat{\mathscr{E}}$.

Let $\tilde{\mathscr{E}} \subseteq \mathscr{E}$ denote the closed U-invariant subspace generated by $\mathscr{E}_0$ and $\tilde{\mathscr{E}}_+ := \tilde{\mathscr{E}} \cap \mathscr{E}_+$. Then $\theta U_t \mathscr{E}_0 = U_{-t} \theta \mathscr{E}_0 = U_{-t} \mathscr{E}_0$ implies that $\tilde{\mathscr{E}}$ is θ-invariant. Therefore $U_t' := U_t|_{\tilde{\mathscr{E}}}$ is a reflection positive unitary one-parameter group on $(\tilde{\mathscr{E}}, \tilde{\mathscr{E}}_+, \theta|_{\tilde{\mathscr{E}}})$. Since $q|_{\tilde{\mathscr{E}}_+'}$ has dense range, all the relevant data is contained in $\tilde{\mathscr{E}}$. It is therefore natural to assume that $\mathscr{E}_0$ is U-cyclic in $\mathscr{E}$ whenever $q(\mathscr{E}_0) = \widehat{\mathscr{E}_0}$ is cyclic in $\widehat{\mathscr{E}}$.

The following proposition shows that the OS transform is compatible with the passage to the space of fixed points.

Proposition 4.2.6 (OS transform commutes with reduction) *Let* $(U_t)_{t \in \mathbb{R}}$ *be a reflection positive unitary one-parameter group on* $(\mathscr{E}, \mathscr{E}_+, \theta)$. *Suppose that* $\mathscr{E}_+$ *is* U-*cyclic and that* $(\widehat{U}_t)_{t \geq 0}$ *is the corresponding one-parameter semigroup of contractions on* $\widehat{\mathscr{E}}$. *Let* $\mathscr{E}_{\text{fix}}$ *denote the subspace of elements fixed under all* U_t *and* $\widehat{\mathscr{E}}_{\text{fix}}$ *the subspace of fixed points for the semigroup* $(\widehat{U}_t)_{t > 0}$. *Then the following assertions hold:*

(a) $\mathscr{E}_{\text{fix}} \subseteq \mathscr{E}_0$, the space of θ-fixed points in $\mathscr{E}_+$.
(b) The map $q|_{\mathscr{E}_{\text{fix}}} : \mathscr{E}_{\text{fix}} \to \widehat{\mathscr{E}}_{\text{fix}}, v \mapsto \widehat{v}$ is a unitary isomorphism.
(c) $\mathscr{E}_{\text{fix}} = \mathscr{E}_\infty := \bigcap_{t>0} U_t \mathscr{E}_+$.

Proof (a) We write $P \colon \mathscr{E} \to \mathscr{E}_{\text{fix}}$ for the orthogonal projection onto the subspace of U-fixed points in $\mathscr{E}$. Then

$$\lim_{N \to \infty} \frac{1}{N} \int_0^N U_t \, dt = P$$

holds in the strong operator topology ([EN00, Corollary V.4.6]). For any $v \in U_s \mathscr{E}_+$ and $s \in \mathbb{R}$, there exists a $T > 0$ with $U_t v \in \mathscr{E}_+$ for $t > T$. Since

$$\lim_{N \to \infty} \frac{1}{N} \int_0^T U_t \, dt = 0,$$

we obtain $Pv \in \mathscr{E}_+$ for every $v \in U_s \mathscr{E}_+$. As $\mathscr{E}_+$ is U-cyclic, we thus obtain $\mathscr{E}_{\text{fix}} = P\mathscr{E} \subseteq \mathscr{E}_+$. Since $\theta U_t \theta = U_{-t}$ for $t \in \mathbb{R}$, the subspace $\mathscr{E}_{\text{fix}}$ is θ-invariant. Now the θ-positivity of $\mathscr{E}_+$ implies that $\theta|_{\mathscr{E}_{\text{fix}}} \geq 0$, and thus $\mathscr{E}_{\text{fix}} \subseteq \mathscr{E}^\theta$.

(b) Since P commutes with θ, Lemma 3.1.2(c) shows that P defines a hermitian contraction $\widehat{P} \colon \widehat{\mathscr{E}} \to \widehat{\mathscr{E}}$ with $\widehat{P}\widehat{v} = \widehat{Pv}$ for $v \in \mathscr{E}_+$. For $v, w \in \mathscr{E}_+$, we obtain

$$\lim_{N \to \infty} \frac{1}{N} \int_0^N \langle \widehat{v}, \widehat{U}_t \widehat{w} \rangle \, dt = \lim_{N \to \infty} \frac{1}{N} \int_0^N \langle \theta v, U_t w \rangle \, dt = \langle \theta v, P w \rangle = \langle \widehat{v}, \widehat{P}\widehat{w} \rangle.$$

Hence [EN00, Corollary V.4.6] implies that $\widehat{P}$ is the orthogonal projection onto $\widehat{\mathscr{E}}_{\text{fix}}$.

Let $q \colon \mathscr{E}_+ \to \widehat{\mathscr{E}}, v \mapsto \widehat{v}$, denote the canonical projection onto $\widehat{\mathscr{E}}$. Then $q \circ P = \widehat{P} \circ q$ implies that $q(\mathscr{E}_{\text{fix}}) = q(P\mathscr{E}_+) = \widehat{P}q(\mathscr{E}_+)$, and hence that $q(\mathscr{E}_{\text{fix}}) \subseteq \widehat{\mathscr{E}}_{\text{fix}}$ is a dense subspace. On the other hand, $\mathscr{E}_{\text{fix}} \subseteq \mathscr{E}_0$ implies that $q|_{\mathscr{E}_{\text{fix}}}$ is isometric, hence a unitary isomorphism onto $\widehat{\mathscr{E}}_{\text{fix}}$.

(c) The subspace $\mathscr{E}_\infty$ is closed and it is easily seen to be invariant under U. Therefore $\mathscr{F} := \mathscr{E}_\infty + \theta \mathscr{E}_\infty$ is invariant under U and θ, so that we obtain a reflection positive unitary one-parameter group $V_t := U_t|_{\mathscr{F}}$ on $(\mathscr{F}, \mathscr{F}_+, \theta|_{\mathscr{F}})$ with $\mathscr{F}_+ := \mathscr{E}_\infty$, satisfying $V_t \mathscr{F}_+ = \mathscr{F}_+$ for every $t > 0$. Now Lemma 3.1.2(d) leads to $\widehat{v}_t = \widehat{V}_{t/2}\widehat{v}_{t/2} = 1$ for every $t > 0$. Therefore $\widehat{\mathscr{F}} \subseteq \widehat{\mathscr{E}}_{\text{fix}}$, and (b) implies that $\widehat{\mathscr{F}} \subseteq q(\mathscr{E}_{\text{fix}})$, so that $\mathscr{E}_\infty = \mathscr{F}_+ \subseteq \mathscr{E}_{\text{fix}} + \mathscr{N}$.

Since the elements of $\mathscr{E}_{\text{fix}}$ are θ-fixed and $\mathscr{N} = \mathscr{E}_+ \cap \theta(\mathscr{E}_+)^\perp$, we have $\mathscr{N} \perp \mathscr{E}_{\text{fix}}$. From $\mathscr{E}_{\text{fix}} \subseteq \mathscr{E}_\infty$ it thus follows that $\mathscr{E}_\infty = \mathscr{E}_{\text{fix}} \oplus (\mathscr{N} \cap \mathscr{E}_\infty)$ is a U-invariant orthogonal decomposition. As $\mathscr{N} \cap \mathscr{E}_\infty$ is orthogonal to the U-cyclic subspace $\theta(\mathscr{E}_+)$, it must be zero, and this shows that $\mathscr{E}_\infty = \mathscr{E}_{\text{fix}}$. $\qquad\square$

Remark 4.2.7 Let $\mathscr{E}^1 := \mathscr{E}_{\text{fix}}^\perp$ in the context of Proposition 4.2.6. Then the reflection positive one-parameter group is adapted to the orthogonal decomposition $\mathscr{E} = \mathscr{E}_{\text{fix}} \oplus \mathscr{E}^1$:

$$\mathscr{E}_+ = \mathscr{E}_{\text{fix}} \oplus \mathscr{E}_+^1, \quad \theta = \mathbf{1} \oplus \theta_1, \quad U_t = \mathbf{1} \oplus U_t^1$$

with respect to the obvious notation. The data corresponding to $\mathscr{E}_{\text{fix}}$ is trivial and the one-parameter group $(U_t^1)_{t \in \mathbb{R}}$ on $(\mathscr{E}^1, \mathscr{E}_+^1, \theta)$ has the additional property that $\mathscr{E}_{\text{fix}}^1 = \{0\}$. We also have that $\widehat{\mathscr{E}} \cong \widehat{\mathscr{E}}_{\text{fix}} \oplus \widehat{\mathscr{E}}_1$.

4.3 Reflection Positive Operator-Valued Functions

We start with a characterization of continuous reflection positive functions for the symmetric semigroup $(G, \tau, S) = (\mathbb{R}, -\operatorname{id}_{\mathbb{R}}, \mathbb{R}_+)$. This is motivated by the GNS construction in Theorem 3.4.5.

Proposition 4.3.1 (Integral representation of reflection positive functions) *Let $\mathscr{F}$ be a Hilbert space and $\varphi \colon \mathbb{R} \to B(\mathscr{F})$ be strongly continuous. Then φ is reflection positive if and only if there exists a finite $\operatorname{Herm}(\mathscr{F})_+$-valued Borel measure Q on $[0, \infty)$ such that*

$$\varphi(x) = \int_0^\infty e^{-\lambda |x|} \, dQ(\lambda). \tag{4.6}$$

Proof Suppose first that φ is reflection positive for $(\mathbb{R}, \mathbb{R}_+, -\operatorname{id})$ and consider the additive unital semigroup $S := ([0, \infty), +)$. Then $\varphi_S := \varphi|_S$ is positive definite with respect to the trivial involution and corresponds to a contraction representation of S because $|\langle \xi, \varphi(s)\xi \rangle| \le \langle \xi, \varphi(e)\xi \rangle$ holds for the positive definite functions $\varphi^{\xi,\xi}(x) := \langle \xi, \varphi(x)\xi \rangle, \xi \in \mathscr{F}$ ([Nel64, Corollary III.1.20(ii)]). Using (A.6) in Example A.1.8 to write $\varphi(s) = \operatorname{ev}_0 \circ U_s^\varphi \circ \operatorname{ev}_0^*$ for the GNS representation $(U^\varphi, \mathscr{H}_\varphi)$ of S and representing U^φ by a spectral measure P on $[0, \infty)$ as

$$U_s^\varphi = \int_0^\infty e^{-\lambda s} \, dP(\lambda)$$

(here we use that the operators U_s^φ are contractions), we obtain the desired integral representation of φ with $Q := \operatorname{ev}_0 \circ P(\cdot) \circ \operatorname{ev}_0^*$. Now (4.6) follows from the fact that $\varphi(-x) = \varphi(x)^* = \varphi(x)$ holds for $x \ge 0$.

For the converse, we assume that φ has an integral representation as in (4.6). This immediately implies that $\varphi|_S$ is positive definite on S for the involution $s^\sharp = s$ and that φ is continuous ([Ne18b, Proposition II.11]). To show that φ is positive definite, we first recall from Example 2.4.3 that

$$e^{-\lambda |x|} = \int_{\mathbb{R}} e^{ixy} \frac{1}{\pi} \frac{\lambda}{\lambda^2 + y^2} \, dy.$$

This implies that

$$\varphi(x) = \int_{\mathbb{R}} e^{ixy} \left(\int_0^\infty \frac{1}{\pi} \frac{\lambda}{\lambda^2 + y^2} \, dQ(\lambda) \right) dy,$$

and since $\tilde{Q}(y) := \int_0^\infty \frac{\lambda}{\lambda^2 + y^2} \, dQ(\lambda)$ is an integrable function with values in positive operators, the positive definiteness of φ follows from Theorem A.2.1. $\square$

Specializing Proposition 4.3.1 to $\mathscr{F} = \mathbb{C}$, we obtain the following integral representation (cf. Example 4.1.4(a)):

Corollary 4.3.2 *A continuous function* $\varphi \colon \mathbb{R} \to \mathbb{C}$ *is reflection positive if and only if it has an integral representation of the form*

$$\varphi(x) = \int_0^\infty e^{-\lambda|x|} \, dv(\lambda), \tag{4.7}$$

where v is a finite positive Borel measure on $[0, \infty)$.

We note the following corollary to the first part of the proof of Proposition 4.3.1:

Corollary 4.3.3 *Let $\mathscr{F}$ be a Hilbert space and* $\varphi \colon [0, \infty) \to B(\mathscr{F})$ *be a bounded strongly continuous function which is positive definite on the $*$-semigroup* $([0, \infty),$ id). *Then $\psi(t) := \varphi(|t|)$ is reflection positive for* $(\mathbb{R}, \mathbb{R}_+, -\mathrm{id}_{\mathbb{R}})$.

Definition 4.3.4 (*Minimal unitary dilation*) If $(C_t)_{t \geq 0}$ is a one-parameter semigroup of hermitian contractions on the Hilbert space $\mathscr{F}$ and $\psi(t) := C_{|t|}$ is the corresponding reflection positive function from Corollary 4.3.3, then the unitary representation U^ψ of $\mathbb{R}$ on the reproducing kernel Hilbert space $\mathscr{H}_\psi$ (Theorem A.1.6) is called the *minimal unitary dilation* of C.

As $\psi(0) = \mathbf{1}$, the space $\mathscr{F}$ may be considered as a subspace of $\mathscr{H}_\psi$ and the orthogonal projection $P \colon \mathscr{H}_\psi \to \mathscr{F}$ satisfies

$$\varphi(s) = P U_s^\psi P^* \quad \text{for} \quad s \geq 0. \tag{4.8}$$

For a detailed account on unitary dilations of semigroups, we refer to [SzN10]; see in particular Proposition 4.3.5 below.

From Proposition 3.4.9 we now derive that $(C_t)_{t \geq 0}$ has a canonical euclidean realization of Markov type in the sense of Definition 3.3.4, but this euclidean realization is rather large as we shall see in Example 4.3.8.

Proposition 4.3.5 *For every strongly continuous one-parameter semigroup* $(C_t)_{t \geq 0}$ *of hermitian contractions on a Hilbert space $\mathscr{H}$, there exists a euclidean realization* $(U_t)_{t \in \mathbb{R}}$ *of Markov type on* $(\mathscr{E}, \mathscr{E}_+, \theta)$ *with $\mathscr{E}_0$ cyclic in $\mathscr{E}$ and $\mathscr{E}_+ = [\![U_{\mathbb{R}_+} \mathscr{E}_0]\!]$. Any realization with these two properties is equivalent to the minimal unitary dilation obtained by the $B(\mathscr{H})$-valued positive definite function $\psi(t) := C_{|t|}$ on $\mathbb{R}$.*

We now develop a concrete picture of the minimal unitary dilation of a contraction semigroup.

Example 4.3.6 Let $(C_t)_{t\geq 0}$ be a continuous semigroup of hermitian contractions on $\mathcal{H}$. We write $C_t = e^{-tH}$ for a selfadjoint non-negative operator $H \geq 0$. To exclude trivialities, we assume that ker $H = 0$, so that the spectral measure E of H is supported by $(0, \infty)$ and we have $H = \int_0^\infty x\, dE(x)$.

(a) On the Hilbert space $\mathscr{E} := L^2(\mathbb{R}, \mathcal{H})$ we consider the unitary one-parameter group given by $(U_t f)(p) = e^{itp} f(p)$. We claim that $(U, \mathscr{E})$ is equivalent to the minimal unitary dilation of $(C_t)_{t\geq 0}$. To verify this claim, we consider the map

$$j: \mathcal{H} \to \mathscr{E}, \qquad j(\xi)(p) := \frac{1}{\sqrt{\pi}} H^{1/2}(H + ip\mathbf{1})^{-1}\xi \quad \text{for} \quad p \neq 0.$$

For $p \neq 0$, the operators $H^{1/2}(H \pm ip\mathbf{1})^{-1}$ are bounded. For $\xi, \eta \in \mathcal{H}$, we have

$$\begin{aligned}
\langle j(\xi), U_t j(\eta) \rangle &= \frac{1}{\pi} \int_{\mathbb{R}} \langle H^{1/2}(H + ip\mathbf{1})^{-1}\xi, e^{itp} H^{1/2}(H + ip\mathbf{1})^{-1}\eta \rangle\, dp \\
&= \frac{1}{\pi} \int_{\mathbb{R}} \int_0^\infty \frac{x}{x^2 + p^2} e^{itp}\, dE^{\xi,\eta}(x)\, dp \quad \text{for} \quad E^{\xi,\eta} = \langle \xi, E(\cdot)\eta \rangle \\
&= \int_0^\infty \left(\int_{\mathbb{R}} \frac{1}{\pi} \frac{x}{x^2 + p^2} e^{itp}\, dp \right) dE^{\xi,\eta}(x) \\
&= \int_0^\infty e^{-x|t|}\, dE^{\xi,\eta}(x) = \langle \xi, e^{-|t|H}\eta \rangle = \langle \xi, C_{|t|}\eta \rangle.
\end{aligned}$$

For $t = 0$, this calculation implies that j is isometric onto the subspace $\mathscr{E}_0 := j(\mathcal{H})$ of $\mathscr{E}$ and that the representation on the subspace $\tilde{\mathscr{E}} := [\![U_{\mathbb{R}}\mathscr{E}_0]\!]$ is equivalent to the GNS representation $(U^\psi, \mathcal{H}_\psi)$ for $\psi(t) := C_{|t|}$, hence the minimal unitary dilation.

To verify our claim, it remains to show that $\mathscr{E} = \tilde{\mathscr{E}}$, i.e., that $\mathscr{E}_0$ is U-cyclic. Since $\mathcal{H}$ is generated by the C-invariant spectral subspaces $\mathcal{H}_{a,b} := E([a, b])\mathcal{H}$, $0 < a < b$, of H, it suffices to argue that $\tilde{\mathscr{E}}$ contains all subspaces $L^2(\mathbb{R}, \mathcal{H}_{a,b})$. Multiplication with $m(p) := H^{1/2}(H + ip\mathbf{1})^{-1}$ defines for $c < d$ a bounded invertible operator $(Af)(p) = m(p)f(p)$ on each subspace $L^2([c, d], \mathcal{H}_{a,b})$ which commutes with U. Hence

$$L^2([c, d], \mathcal{H}_{a,b}) = AL^2([c, d], \mathcal{H}_{a,b}) = [\![L^\infty([c, d])j(\mathcal{H}_{a,b})]\!] \subseteq [\![U_{\mathbb{R}}j(\mathcal{H}_{a,b})]\!]$$

follows from $U_{\mathbb{R}}'' = L^\infty(\mathbb{R})$ and therefore $\mathscr{E}_0$ is U-cyclic and $\tilde{\mathscr{E}} = \mathscr{E}$.

(b) Now we determine the subspace $\mathscr{E}_+ = [\![U_{\mathbb{R}_+}\mathscr{E}_0]\!]$ and the involution θ. From $\mathscr{E}_0 \subseteq \mathscr{E}^\theta, \theta U_t \theta = U_{-t}$ and the cyclicity of $\mathscr{E}_0$, we immediately obtain

$$(\theta f)(p) = \frac{H - ip\mathbf{1}}{H + ip\mathbf{1}} f(-p) := (H - ip\mathbf{1})(H + ip\mathbf{1})^{-1} f(-p),$$

so that U is reflection positive on $(\mathscr{E}, \mathscr{E}_+, \theta)$. The Markov property follows from the multiplicativity of $\psi(t) = C_{|t|}$ for $t \geq 0$ (Proposition 3.4.9).

Next we show that $\mathscr{E}_+ \subseteq \mathscr{E}$ coincides with $L^2_+(\mathbb{R}, \mathscr{H}) := \mathscr{F} L^2(\mathbb{R}_+, \mathscr{H})$, where $\mathscr{F}$ is the Fourier transform. Writing $\mathbb{C}_+ := \mathbb{R}_+ + i\mathbb{R}$ for the right half plane, the map

$$\frac{1}{\sqrt{2\pi}} \mathscr{L} : L^2(\mathbb{R}_+, \mathscr{H}) \to \mathscr{O}(\mathbb{C}_+, \mathscr{H}), \quad \mathscr{L}(f)(z) = \int_0^\infty e^{-zx} f(x)\, dx \quad (4.9)$$

is isometric onto the $\mathscr{H}$-valued *Hardy space* $H^2(\mathbb{C}_+, \mathscr{H})$ with the norm

$$\|f\|^2 = \lim_{z \to 0_+} \int_{\mathbb{R}} \|f(z + ip)\|^2\, dp.$$

Its reproducing kernel $Q(z, w) \in B(\mathscr{H})$ is given by $Q(z, w) = \frac{1}{2\pi} \frac{1}{z+w} \mathbf{1}$ because the functions $Q_{z,\xi} := (2\pi)^{-1/2} e_{-\bar{z}} \chi_{\mathbb{R}_+} \xi \in L^2(\mathbb{R}_+, \mathscr{H})$ with $e_z(x) = e^{zx}$ and $\xi \in \mathscr{H}$ satisfy

$$\langle Q_{z,\xi}, Q_{w,\eta} \rangle = \frac{1}{2\pi} \int_0^\infty e^{-x(z+\bar{w})} \langle \xi, \eta \rangle\, dx = \frac{1}{2\pi} \frac{\langle \xi, \eta \rangle}{z + \bar{w}}. \quad (4.10)$$

That $\mathscr{E}_0$ is contained in $L^2_+(\mathbb{R}, \mathscr{H})$ follows from the following calculation for $\operatorname{Re} z \geq 0$ (and evaluating in $z = ip$), where we put $E^\xi := E(\cdot)\xi$:

$$H^{1/2}(H + z\mathbf{1})^{-1}\xi = \int_0^\infty \frac{x^{1/2}}{x + z} dE^\xi(x) = \int_0^\infty \int_0^\infty e^{-\lambda z} e^{-\lambda x} x^{1/2}\, d\lambda\, dE^\xi(x)$$
$$= \int_0^\infty \int_0^\infty e^{-\lambda z} e^{-\lambda x} x^{1/2} dE^\xi(x)\, d\lambda = \int_0^\infty e^{-\lambda z} \left(e^{-\lambda H} H^{1/2}\xi \right) d\lambda.$$

Now $\mathscr{E}_+ = [\![U_{\mathbb{R}_+} \mathscr{E}_0]\!] \subseteq L^2_+(\mathbb{R}, \mathscr{H})$ follows from the invariance of $L^2_+(\mathbb{R}, \mathscr{H})$ under the operators $U_t = \mathscr{F} \circ V_t \circ \mathscr{F}^{-1}$ for $t \geq 0$, where $(V_t f)(x) = f(x - t)$. In view of the maximal θ-positivity of $\mathscr{E}_+$ (Lemma 2.3.2), equality will follow if the Hardy space is θ-positive. This is verified as follows. The functions $f_{z,\xi}(p) := \frac{1}{\bar{z}+ip}\xi = \int_0^\infty e^{-x(\bar{z}+ip)}\xi\, dx$, $\operatorname{Re} z > 0$, $\xi \in \mathscr{H}$, generate $L^2_+(\mathbb{R}, \mathscr{H})$. We have

$$\langle f_{z,\xi}, \theta f_{w,\eta} \rangle = \int_{\mathbb{R}} \frac{\langle \xi, (H - ip\mathbf{1})(H + ip\mathbf{1})^{-1}\eta \rangle}{(z - ip)(\bar{w} - ip)}\, dp \quad \text{for} \quad \operatorname{Re} z, \operatorname{Re} w > 0.$$
$$(4.11)$$

Since the function

$$G(\zeta) := \frac{\langle \xi, (H - i\zeta\mathbf{1})(H + i\zeta\mathbf{1})^{-1}\eta \rangle}{(z - i\zeta)(\bar{w} - i\zeta)} = -\frac{\langle \xi, (H - i\zeta\mathbf{1})(H + i\zeta\mathbf{1})^{-1}\eta \rangle}{(\zeta + iz)(\zeta + i\bar{w})}$$

is meromorphic in the lower half plane $\{\operatorname{Im} \zeta < 0\}$ with poles in $-iz$ and $-i\bar{w}$ and $\lim_{\zeta \to \infty} |\zeta G(\zeta)| = 0$, the Residue Theorem, applied to negatively oriented paths in the lower half plane which lead to winding number -1, yields for $z \neq \bar{w}$:

$$\langle f_{z,\xi}, \theta f_{w,\eta} \rangle = 2\pi i(-\operatorname{Res}_{-iz}(G) - \operatorname{Res}_{-iw}(G)),$$

where

$$\operatorname{Res}_{-iz}(G) = -\frac{\langle \xi, (H - z\mathbf{1})(H + z\mathbf{1})^{-1}\eta \rangle}{-iz + i\overline{w}} = -i\frac{\langle \xi, (H - z\mathbf{1})(H + z\mathbf{1})^{-1}\eta \rangle}{z - \overline{w}}$$

and

$$\operatorname{Res}_{-i\overline{w}}(G) = -\frac{\langle \xi, (H - \overline{w}\mathbf{1})(H + \overline{w}\mathbf{1})^{-1}\eta \rangle}{-i\overline{w} + iz} = i\frac{\langle \xi, (H - \overline{w}\mathbf{1})(H + \overline{w}\mathbf{1})^{-1}\eta \rangle}{z - \overline{w}}.$$

We thus arrive at

$$\begin{aligned}
\langle f_{z,\xi}, \theta f_{w,\eta} \rangle &= \frac{2\pi}{z - \overline{w}} \langle \xi, (z\mathbf{1} - H)(H + z\mathbf{1})^{-1} - (\overline{w}\mathbf{1} - H)(H + \overline{w}\mathbf{1})^{-1}\eta \rangle \\
&= 4\pi \langle \xi, (H + z\mathbf{1})^{-1}H(H + \overline{w}\mathbf{1})^{-1}\eta \rangle \\
&= 4\pi \langle H^{1/2}(H + \overline{z}\mathbf{1})^{-1}\xi, H^{1/2}(H + \overline{w}\mathbf{1})^{-1}\eta \rangle,
\end{aligned}$$

which obviously is a positive definite kernel on $\mathbb{C}_+ \times \mathcal{H}$, and therefore $L^2_+(\mathbb{R}, \mathcal{H})$ is θ-positive.

(c) Finally we note that the map

$$(Tf)(p) := \sqrt{\pi} H^{-1/2}(H + ip\mathbf{1})f(p)$$

maps $L^2(\mathbb{R}, \mathcal{H})$ unitarily onto the space $\tilde{\mathcal{E}}$ of all $\mathcal{H}$-valued L^2-functions with respect to the norm given by

$$\|f\|^2_H := \frac{1}{\pi} \int_{\mathbb{R}} \langle f(p), H(H^2 + p^2)^{-1}f(p) \rangle \, dp.$$

The operator T intertwines U with the representation $(\tilde{U}_t f)(p) = e^{itp}f(p)$ and the involution θ with $(\tilde{\theta}f)(p) := f(-p)$.

Example 4.3.7 We take a closer look at the Hardy space $H^2(\mathbb{C}_+, \mathcal{H}) = \mathcal{H}_Q \subseteq \mathcal{O}(\mathbb{C}_+, \mathcal{H})$ with the reproducing kernel $Q(z, w) = \frac{1}{z+\overline{w}}\mathbf{1}$ introduced in Example 4.3.6(b). For simplicity we omit the factor $\frac{1}{2\pi}$, so that the Laplace transform $\mathcal{L}: L^2(\mathbb{R}_+, \mathcal{H}) \to H^2(\mathbb{C}_+, \mathcal{H})$ is unitary.

(a) For the translation action $(V_t f)(x) = f(x - t)$ on $L^2(\mathbb{R}, \mathcal{H})$ we have for $t \geq 0$ and $f \in L^2(\mathbb{R}_+, \mathcal{H})$:

$$\mathcal{L}(V_t f)(z) = \int_t^\infty e^{-xz}f(x - t)\, dt = \int_0^\infty e^{-(x+t)z}f(x)\, dt = e^{-tz}\mathcal{L}(f)(z),$$

(4.12)

so that the subsemigroup $\mathbb{R}_+ \subseteq \mathbb{R}$ acts on $H^2(\mathbb{C}_+, \mathcal{H})$ by multiplication $(U_t f)(z) = e^{-tz} f(z), t \geq 0$. This action is isometric because the boundary values of $e_{-t}(z) = e^{-tz}$ on $i\mathbb{R}$ have absolute value 1.

(b) We now specialize to the scalar case where $\mathcal{H} = \mathbb{C}$ and $C_t = e^{-t\lambda}$ for some $\lambda > 0$. Then Example 4.3.6(a) shows that the subspace $\mathcal{E}_0$ of $\mathcal{E}_+ \cong H^2(\mathbb{C}_+, \mathbb{C})$ is generated by the function $Q_\lambda(z) = \frac{1}{\lambda+z}$ with $\|Q_\lambda\|^2 = Q(\lambda, \lambda) = \frac{1}{2\lambda}$. Therefore the quotient map $q \colon \mathcal{E}_+ \cong H^2(\mathbb{C}_+, \mathbb{C}) \to \widehat{\mathcal{E}} \cong \mathbb{C}$ is given by evaluation:

$$q(f) = \sqrt{2\lambda}\langle Q_\lambda, f \rangle = \sqrt{2\lambda} f(\lambda).$$

This formula also shows immediately that

$$q(U_t f) = e^{-t\lambda} q(f) \quad \text{for} \quad f \in H^2(\mathbb{C}_+, \mathbb{C}), t \geq 0.$$

(c) (Anti-unitary involutions) On the space $\mathcal{E} = L^2(\mathbb{R})$, the conjugation

$$(Jf)(p) := \overline{f(-p)}$$

commutes with $(U_t)_{t \in \mathbb{R}}$ and satisfies $J\mathcal{E}_+ = \mathcal{E}_+$ and $JQ_z = Q_{\bar{z}}$ for $\operatorname{Re} z \geq 0$. Therefore it induces on the Hardy space $H^2(\mathbb{C}_+, \mathbb{C})$ the conjugation given by

$$(Jf)(z) = \overline{f(\bar{z})} \quad \text{for} \quad f \in \mathcal{H}_Q.$$

We also observe that $J\theta = \theta J$ on $L^2(\mathbb{R})$, and since $JQ_\lambda = Q_\lambda$, it induces on $\widehat{\mathcal{E}} \cong \mathbb{C}$ the involution given by complex conjugation.

Example 4.3.8 (Cyclic contraction semigroups) For a σ-finite measure space $(X, \mathfrak{S}, \rho)$ and a measurable function $h \colon X \to [0, \infty)$, consider on $\mathcal{H} := L^2(X, \rho)$ the hermitian contraction semigroup $C_t f = e^{-th} f$. By the Spectral Theorem, all cyclic contraction semigroups can be represented this way with a finite measure ρ on $X = [0, \infty)$ and $h(\lambda) = \lambda$, so that 1 is a cyclic vector.

(a) With Example 4.3.6(c), we obtain a euclidean realization of Markov type (and, in general, infinite multiplicity) by

$$\mathcal{E} = L^2(\mathbb{R}^\times \times X, \zeta), \quad d\zeta(x, \lambda) = \left(\frac{1}{\pi} \frac{h(\lambda)}{h(\lambda)^2 + x^2} dx\right) d\rho(\lambda),$$
$$(U_t f)(x, \lambda) = e^{itx} f(x, \lambda), \qquad (\theta f)(x, \lambda) := f(-x, \lambda).$$

Here $\mathcal{E}_0 \cong L^2(X, \rho)$ is the subspace of functions $f(x, \lambda) = f(\lambda)$ not depending on x and this subspace is U-cyclic in $\mathcal{E}$.

(b) For a finite measure ρ on $X = [0, \infty) = \mathbb{R}_{\geq 0}$ and $h(\lambda) = \lambda$, a multiplicity free euclidean realizations can be obtained as follows. The projection $\operatorname{pr} \colon \mathbb{R} \times \mathbb{R}_{\geq 0} \to \mathbb{R}, \operatorname{pr}(x, \lambda) := x$ maps the measure ζ to the measure $\nu := \operatorname{pr}_* \zeta$ given by

$$dv(x) = \frac{1}{\pi} \left(\int_0^\infty \frac{\lambda}{\lambda^2 + x^2} \, d\rho(\lambda) \right) dx. \tag{4.13}$$

Then $\mathscr{F} := L^2(\mathbb{R}, \nu)$ can be identified with the U-invariant subspace of $\mathscr{E}$ consisting of functions not depending on λ. For $\mathscr{F}_+ := \mathscr{F} \cap \mathscr{E}_+$, we obtain on $(\mathscr{F}, \mathscr{F}_+, \theta)$ a reflection positive one-parameter group $(V_t)_{t \in \mathbb{R}}$ by restriction. Now $\mathscr{F}_0 = \mathscr{F}_+^\theta = \mathscr{E}_0 \cap \mathscr{F} = \mathbb{C}1$ is the space of constant functions. Here $1 \in \mathscr{F}_0$ is a cyclic vector corresponding to the reflection positive function

$$\varphi(t) := \langle 1, V_t 1 \rangle = \int_{\mathbb{R}} e^{-itx} \, d\nu(x) = \int_0^\infty e^{-\lambda|t|} \, d\rho(\lambda)$$

(Example 2.4.3(a)). Its restriction to $\mathbb{R}_+$ leads to a GNS representation equivalent to the multiplication representation of $\mathbb{R}_+$ on $\mathscr{H} = L^2(\mathbb{R}_{\geq 0}, \rho)$ given by C, so that $(V, \mathscr{F}, \mathscr{F}_+, \theta)$ also is a euclidean realization of $(C, \mathscr{H})$, but not of Markov type if ρ is not a point measure.

In both cases, the subspaces $\mathscr{E}_0$ and $\mathscr{F}_0$, respectively, are cyclic, but in the first case the Markov condition $q(\mathscr{E}_0) = \widehat{\mathscr{E}}$ holds, whereas in the second case $\mathscr{F}_0 = \mathbb{C}1$ is one-dimensional.

(c) For $\dim \mathscr{E}_0 = 1$ and $C_t = e^{-t\lambda}$, $\lambda \geq 0$, the minimal dilation $\varphi(t) = e^{-\lambda|t|}$ from (a) leads to the Hilbert space $\mathscr{E} \cong L^2(\mathbb{R}, \frac{\lambda}{\pi} \frac{dx}{\lambda^2 + x^2})$ because $\rho = \delta_\lambda$ is a point measure. In this case the realizations in (a) and (b) coincide.

4.4 A Connection to Lax–Phillips Scattering Theory

One parameter groups and reflection positivity are closely related to the Sinai/Lax-Phillips scattering theory and translation invariant subspaces ([LP64, LP67, LP81, Sin61]). In short, this theory says that every unitary representation of $\mathbb{R}$ on a Hilbert space $\mathscr{E}$ satisfying some simple conditions stated below can be realized by translations in $L^2(\mathbb{R}, \mathscr{M})$ for some multiplicity Hilbert space $\mathscr{M}$.

Let $(U, \mathscr{E})$ be a unitary representation of $\mathbb{R}$. A closed subspace $\mathscr{E}_+ \subset \mathscr{E}$ is called *outgoing* if

(LP1) $\mathscr{E}_+$ is invariant under U_t, $t > 0$,

(LP2) $\mathscr{E}_\infty := \bigcap_{t>0} U_t \mathscr{E}_+ = \{0\}$,

(LP3) $\bigcup_{t<0} U_t \mathscr{E}_+$ is dense in $\mathscr{E}$.

The following theorem is classical ([LP64, Theorem 1]):

Theorem 4.4.1 (Lax–Phillips Representation Theorem) *If $\mathscr{E}_+$ is outgoing for $(U, \mathscr{E})$, then there exists a Hilbert space $\mathscr{M}$ such that $\mathscr{E} \simeq L^2(\mathbb{R}, \mathscr{M})$, $\mathscr{E}_+ \simeq$*

$L^2([0, \infty), \mathscr{M})$, and U is represented by translations $(U_t f)(x) = f(x - t)$. This representation is unique up to isomorphism of $\mathscr{M}$.

This realization of $(U, \mathscr{E})$ is called the *outgoing realization* of U.

Proof [1] For $t \in \mathbb{R}$, we put $\mathscr{E}_t := U_t \mathscr{E}_+$ and write E_t for the corresponding projection. Then our assumptions imply that

$$\lim_{t \to \infty} E_t = \{0\} \quad \text{and} \quad \lim_{t \to -\infty} E_t = \mathbf{1}$$

in the strong operator topology. We further have $\lim_{s \to t_-} E_s = E_t$. In fact, $\mathscr{E}_t \subseteq \mathscr{E}_s$ for $s < t$ implies that $E_t^- := \lim_{s \to t_-} E_s \geq E_t$ exists. If, conversely, $v \in E_t^- \mathscr{E} = \bigcap_{s<t} \mathscr{E}_s$, we have for every $h > 0$ that $U_h v \in \mathscr{E}_t$ and thus $v \in \mathscr{E}_t$ by closedness. Thus $E_t^- \leq E_t$, and therefore $E_t^- = E_t$. We conclude that the family $(E_t)_{t \in \mathbb{R}}$ defines a Stieltjes spectral measure P on $\mathfrak{B}(\mathbb{R})$ such that $P([t, \infty)) = E_t$ for $t \in \mathbb{R}$.

Consider the unitary one-parameter group defined by $V_s := \int_{\mathbb{R}} e^{isx} \, dP(x)$. Then $U_t E_x = E_{x+t}$ shows that $U_t P([x, \infty)) = P([x + t, \infty))$ for $x, t \in \mathbb{R}$, and therefore $U_t P([a, b]) = P([t + a, t + b])$ for $a < b$. This implies

$$U_t V_s U_{-t} = \int_{\mathbb{R}} e^{isx} dP(x + t) = \int_{\mathbb{R}} e^{is(x-t)} dP(x) = e^{-ist} V_s.$$

Therefore we obtain a unitary representation of the Heisenberg group $\text{Heis}(\mathbb{R}^2) = \mathbb{T} \times \mathbb{R}^2$ with the product

$$(z, s, t)(z', s', t') = (zz' e^{-its'}, s + s', t + t') \quad \text{by} \quad \pi(z, s, t) := z V_s U_t.$$

Now the assertion follows from the Stone–von Neumann Theorem ([Nel64, Theorem X.3.1]).   □

We now connect the Lax–Phillips construction to the dilation process. The following proposition is an obvious consequence of the Lax–Phillips Theorem 4.4.1 and Proposition 4.2.6.

Proposition 4.4.2 *Let $(U_t)_{t \in \mathbb{R}}$ be a reflection positive unitary one-parameter group on $(\mathscr{E}, \mathscr{E}_+, \theta)$ for which $\mathscr{E}_+$ is cyclic and $\mathscr{E}_{\text{fix}} = \{0\}$. Then $\mathscr{E}_+$ is outgoing, so that $(U, \mathscr{E})$ is unitarily equivalent to the translation representation on $L^2(\mathbb{R}, \mathscr{M})$ for some Hilbert space $\mathscr{M}$. This realization is unique up to isomorphism of $\mathscr{M}$.*

Example 4.4.3 As we have seen in Example 4.3.6, the Fourier transform immediately yields an outgoing realization of the minimal dilation representation of a contraction semigroup $(C_t)_{t \geq 0}$ with trivial fixed points on $\mathscr{H}$ on the space $\mathscr{E} = L^2(\mathbb{R}, \mathscr{H})$.

Remark 4.4.4 Proposition 4.4.2 shows in particular that, up to a direct summand consisting of fixed points, the spectrum of any euclidean realization is all of $\mathbb{R}$ and the representation is a multiple of the translation representation of $\mathbb{R}$ on $L^2(\mathbb{R})$.

---

[1] We thank Bent Ørsted for communicating this short representation theoretic proof.

Proposition 4.4.2 suggests to attempt a classification of reflection positive one-parameter groups in an outgoing realization on $\mathscr{E} = L^2(\mathbb{R}, \mathscr{M})$ with $\mathscr{E}_+ = L^2(\mathbb{R}_+, \mathscr{M})$ by classifying the unitary equivalence classes of corresponding unitary involutions.

The Fourier transform of the subspace $\mathscr{E}_+ = L^2(\mathbb{R}_+, \mathscr{M})$ is the $\mathscr{M}$-valued Hardy space $\tilde{\mathscr{E}}_+ = H^2(\mathbb{C}_+, \mathscr{H}) \subseteq \mathcal{O}(\mathbb{C}_+, \mathscr{M})$ which can be considered as a closed subspace of $\tilde{\mathscr{E}} := L^2(\mathbb{R}, \mathscr{M})$ by the natural boundary-value map. The translation action of $\mathbb{R}$ on $\mathscr{E}$ leads to the multiplication action

$$(\tilde{U}_t f)(z) = e^{-tz} f(z) \quad \text{for} \quad f \in \tilde{\mathscr{E}}_+.$$

To classify reflection positive one-parameter groups (without fixed points) now corresponds to the problem to determine the involutions θ on $\tilde{\mathscr{E}}$ for which $\tilde{\mathscr{E}}_+$ is θ-positive and $\theta \tilde{U}_t \theta = \tilde{U}_{-t}$. Since the commutant of the multiplication action on $\tilde{\mathscr{E}}$ is the von Neumann algebra $L^\infty(\mathbb{R}, B(\mathscr{M}))$, the involution θ must be of the form

$$(\theta f)(p) = m(p) f(-p), \quad \text{where} \quad m \colon \mathbb{R} \to \mathrm{U}(\mathscr{M})$$

is a unitary operator in $L^\infty(\mathbb{R}, B(\mathscr{M}))$, which basically is a measurable map with values in $\mathrm{U}(\mathscr{M}))$ satisfying $m(-p) = m(p)^* = m(p)^{-1}$ almost everywhere. That the Hardy space is θ-positive is by (4.11) equivalent to the positive definiteness of the $B(\mathscr{M})$-valued kernel

$$R(z, w) := \int_{\mathbb{R}} \frac{m(p)}{(z - ip)(\overline{w} - ip)} \, dp. \tag{4.14}$$

In Example 4.3.6, we have seen that this is the case if $m(p) = \frac{H - ip\mathbf{1}}{H + ip\mathbf{1}}$ for a strictly positive operator H on $\mathscr{M}$. This corresponds to the case of reflection positive one-parameter groups of Markov type.

Problem 4.4.5 Characterize unitary-valued functions $m \in L^\infty(\mathbb{R}, B(\mathscr{M}))$ with $m(-p) = m(p)^*$ for which the kernel (4.14) on $\mathbb{C}_+$ is positive definite.

Notes

Section 4.1: The material discussed briefly in Sect. 4.1 is contained in [JNO18]. For recent progress in the local theory of positive definite functions on groups we refer to [JN16, JPT15].

Section 4.2: A version of Corollary 4.3.2 for reflection positivity on the group $\mathbb{Z}$ can be found in [FILS78, Proposition 3.2]. In this context reflection positivity is also analyzed in [JT17].

For the special case where $\varphi(0) = \mathbf{1}$, strongly continuous reflection positive functions $\varphi \colon \mathbb{R} \to B(V)$ are called (OS)-positive covariance functions in [Kl77], and Proposition 4.3.1 specializes to [Kl77, Remark 2.7].

Corollary 4.3.3 can also be found in [SzN10, Theorem I.8.1]. For a detailed account on unitary dilations of semigroups, we refer to [SzN10]; see in particular Proposition 4.3.5.

Chapter 5
Reflection Positivity on the Circle

In this chapter we turn to the close relation between reflection positivity on the circle group $\mathbb{T}$ and the Kubo–Martin–Schwinger (KMS) condition for states of C^*-dynamical systems. Here a crucial point is a pure representation theoretic perspective on the KMS condition formulated as a property of form-valued positive definite functions on $\mathbb{R}$: For $\beta > 0$, we consider the open strip

$$\mathscr{S}_\beta := \{z \in \mathbb{C} \colon 0 < \operatorname{Im} z < \beta\}.$$

For a real vector space V, we say that a positive definite function $\psi \colon \mathbb{R} \to \operatorname{Bil}(V)$ (Definition A.1.5) satisfies the β-*KMS condition* if ψ extends to a pointwise continuous function $\overline{\psi}$ on $\overline{\mathscr{S}_\beta}$ which is pointwise holomorphic on $\mathscr{S}_\beta$ and satisfies $\psi(i\beta + t) = \overline{\psi(t)}$ for $t \in \mathbb{R}$.

The key idea in the classification of positive definite functions satisfying a KMS condition is to relate them to standard (real) subspaces of a (complex) Hilbert space which occur naturally in the modular theory of operator algebras [Lo08]. These are closed real subspaces $V \subseteq \mathscr{H}$ for which $V \cap iV = \{0\}$ and $V + iV$ is dense. Any standard subspace determines a pair (Δ, J) of *modular objects*, where Δ is a positive selfadjoint operator and J an anti-linear involution (a *conjugation*) satisfying $J\Delta J = \Delta^{-1}$. The connection is established by

$$V = \operatorname{Fix}(J\Delta^{1/2}) = \{\xi \in \mathscr{D}(\Delta^{1/2}) \colon J\Delta^{1/2}\xi = \xi\}. \tag{5.1}$$

A key result is the characterization of the KMS condition in terms of standard subspaces (Theorem 5.1.7) which also contains a classification in terms of an integral representation.

For a function ψ satisfying the β-KMS condition, analytic continuation to $\overline{\mathscr{S}_\beta}$ leads to an operator-valued function

$$\tilde{\varphi} \colon [0, \beta] \to B(V_{\mathbb{C}}) \quad \text{by} \quad \langle \xi, \tilde{\varphi}(t)\eta \rangle = \psi(it)(\xi, \eta) \quad \text{for} \quad \xi, \eta \in V.$$

© The Author(s) 2018
K.-H. Neeb and G. Ólafsson, *Reflection Positivity*, SpringerBriefs in Mathematical Physics, https://doi.org/10.1007/978-3-319-94755-6_5

This function satisfies $\tilde{\varphi}(\beta) = \overline{\tilde{\varphi}(0)}$, hence extends uniquely to a weak operator continuous function $\tilde{\varphi} \colon \mathbb{R} \to B(V_{\mathbb{C}})$ satisfying

$$\tilde{\varphi}(t + \beta) = \overline{\tilde{\varphi}(t)} \quad \text{for} \quad t \in \mathbb{R}. \tag{5.2}$$

Here we write complex linear operators on $V_{\mathbb{C}}$ as $A + iB$ with $A, B \in B(V)$ and put $\overline{A + iB} = A - iB$.

Recall the group $\mathbb{R}_\tau = \mathbb{R} \rtimes \{e, \tau\}$ with $\tau(t) = -t$. In Theorem 5.2.3 we show that there exists a natural positive definite function

$$f \colon \mathbb{R}_\tau \to \mathrm{Bil}(V) \quad \text{satisfying} \quad f(t, \tau) = \tilde{\varphi}(t).$$

The function f is 2β-periodic, hence factors through a function on the group

$$\mathbb{T}_{2\beta, \tau} := \mathbb{R}_\tau / \mathbb{Z}2\beta \cong O_2(\mathbb{R})$$

and it is reflection positive for $G = \mathbb{T}_{2\beta}$ and $G_+ = [0, \beta] + 2\beta\mathbb{Z}$. This leads to a natural euclidean realization of the unitary one-parameter group $U_t = \Delta^{-it/\beta}$ associated to ψ. We conclude this section with a description of the GNS representation of $\mathbb{T}_{2\beta, \tau}$ in a natural space of sections of a vector bundle over the circle $\mathbb{R}/\mathbb{Z}\beta$ with two-dimensional fiber on which the scalar product is given by a resolvent of the Laplacian as in Sect. 2.5; see also Sect. 7.4.2.

5.1 Positive Definite Functions Satisfying KMS Conditions

In this section we present a characterization (Theorem 5.1.7) of form-valued positive definite functions on $\mathbb{R}$ satisfying a KMS condition. We also explain how the corresponding representation of $\mathbb{R}$ can be realized in a Hilbert space of holomorphic functions on the strip $\mathscr{S}_{\beta/2}$ with continuous boundary values (Proposition 5.1.11).

We call a function $\psi \colon \overline{\mathscr{S}_\beta} \to \mathrm{Bil}(V)$ *pointwise continuous* if, for all $v, w \in V$, the function $\psi^{v,w}(z) := \psi(z)(v, w)$ is continuous. Moreover, we say that ψ is *pointwise holomorphic in* $\mathscr{S}_\beta$, if, for all $v, w \in V$, the function $\psi^{v,w}|_{\mathscr{S}_\beta}$ is holomorphic. By the Schwarz reflection principle, any pointwise continuous pointwise holomorphic function ψ is uniquely determined by its restriction to $\mathbb{R}$.

Definition 5.1.1 For a real vector space V, we say that a positive definite function $\psi \colon \mathbb{R} \to \mathrm{Bil}(V)$ satisfies the *KMS condition* for $\beta > 0$ if ψ extends to a function $\psi \colon \overline{\mathscr{S}_\beta} \to \mathrm{Bil}(V)$ which is pointwise continuous and pointwise holomorphic on $\mathscr{S}_\beta$, and satisfies

$$\psi(i\beta + t) = \overline{\psi(t)} \quad \text{for} \quad t \in \mathbb{R}. \tag{5.3}$$

Lemma 5.1.2 *Suppose that* $\psi: \mathbb{R} \to \mathrm{Bil}(V)$ *satisfies the KMS condition for* $\beta > 0$. *Then*

$$\psi(i\beta + \bar{z}) = \overline{\psi(z)} = \psi(-\bar{z})^{\top} \quad for \quad z \in \overline{\mathscr{S}_\beta}. \tag{5.4}$$

The function $\varphi: [0, \beta] \to \mathrm{Bil}(V)$, $\varphi(t) := \psi(it)$ *has hermitian values and satisfies*

$$\varphi(\beta - t) = \overline{\varphi(t)} \quad for \quad 0 \leq t \leq \beta. \tag{5.5}$$

It extends to a unique strongly continuous symmetric 2β-*periodic function* $\varphi: \mathbb{R} \to \mathrm{Herm}(V)$ *satisfying*

$$\varphi(\beta + t) = \overline{\varphi(t)} \quad and \quad \varphi(-t) = \varphi(t) \quad for \quad t \in \mathbb{R}.$$

Proof Note that $\psi(-t) = \overline{\psi(t)}^{\top} = \psi(t)^*$ holds for every positive definite function $\psi: \mathbb{R} \to \mathrm{Bil}(V)$. By analytic continuation (resp., the Schwarz Reflection Principle), this leads to the second equality in (5.4). Likewise, condition (5.3) leads to the first equality in (5.4). This in turn implies (5.5), and the remainder is clear. $\square$

To obtain a natural representation of ψ, we now introduce standard subspaces $V \subseteq \mathscr{H}$ and the associated modular objects (Δ, J).

Definition 5.1.3 A closed real subspace V of a complex Hilbert space $\mathscr{H}$ is said to be *standard* if

$$V \cap iV = \{0\} \quad and \quad \overline{V + iV} = \mathscr{H}.$$

For every standard real subspace $V \subseteq \mathscr{H}$, we define an unbounded anti-linear operator

$$S: \mathscr{D}(S) = V + iV \to \mathscr{H}, \quad S(\xi + i\eta) := \xi - i\eta \quad for \quad \xi, \eta \in V.$$

Then S is closed and has a polar decomposition $S = J\Delta^{1/2}$, where J is an anti-unitary involution and Δ a strictly positive selfadjoint operator (cf. [NÓ15b, Lemma 4.2]; see also [BR02, Proposition 2.5.11], [Lo08, Proposition 3.3]). We call (Δ, J) the pair of *modular objects* of V.

Remark 5.1.4 (a) From $S^2 = \mathrm{id}$, it follows that the modular objects (Δ, J) of a standard subspace satisfies the *modular relation*

$$J\Delta J = \Delta^{-1}. \tag{5.6}$$

If, conversely, (Δ, J) is a pair of a strictly positive selfadjoint operator Δ and an anti-unitary involution J satisfying (5.6), then $S := J\Delta^{1/2}$ is an anti-linear involution with $\mathscr{D}(S) = \mathscr{D}(\Delta^{1/2})$ whose fixed point space $\mathrm{Fix}(S)$ is a standard subspace. Thus standard subspaces are in one-to-one correspondence with pairs (Δ, J) satisfying (5.6) (cf. [Lo08, Proposition 3.2] and [NÓ15b, Lemma 4.4]).

(b) As the unitary one-parameter group $(\Delta^{it})_{t \in \mathbb{R}}$ commutes with J and Δ, it leaves the real subspace $V = \mathrm{Fix}(S)$ invariant.

The following proposition ([NÓ15b, Proposition 3.1]) provides various characterizations of unitary one-parameter groups with reflection symmetry. As we shall see below, these are precisely those for which a euclidean realization on $\mathbb{T}_{2\beta,\tau}$ can be obtained by a positive definite function satisfying the β-KMS condition.

Proposition 5.1.5 *For a unitary one-parameter group $(U_t)_{t \in \mathbb{R}}$ on $\mathcal{H}$ with spectral measure $E : \mathfrak{B}(\mathbb{R}) \to B(\mathcal{H})$, the following are equivalent:*

  (i) *There exists an anti-unitary involution J on $\mathcal{H}$ with $J U_t J = U_t$ for $t \in \mathbb{R}$.*
  (ii) *For $\mathcal{H}_\pm := E(\mathbb{R}_\pm^\times)\mathcal{H}$, the unitary one-parameter groups $U_t^+ := U_t|_{\mathcal{H}_+}$ and $U_t^- := U_{-t}|_{\mathcal{H}_-}$ are unitarily equivalent.*
  (iii) *The unitary one-parameter group $(U, \mathcal{H})$ is equivalent to a GNS representation $(U^\psi, \mathcal{H}_\psi)$, where $\psi : \mathbb{R} \to B(V)$ is a symmetric positive definite function.*
  (iv) *There exists a unitary involution θ on $\mathcal{H}$ with $\theta U_t \theta = U_{-t}$ for $t \in \mathbb{R}$.*

Remark 5.1.6 It is easy to see that conditions (i)–(iv) even imply the existence of an extension of U to a representation of the group $\mathbb{R}_\tau \times \{\pm 1\} \cong (\mathbb{R}^\times)_\tau \cong O_{1,1}(\mathbb{R})$ by unitary and anti-unitary operators, where τ is represented by a unitary involution and $(0, -1)$ by a conjugation J. Since any unitary representation is a direct sum of cyclic ones, it suffices to verify our claim in the cyclic case. Under the assumption of Proposition 5.1.5, $(U, \mathcal{H})$ is equivalent to the representation in $\mathcal{H} = L^2(\mathbb{R}, \nu)$ for a finite symmetric measure ν given by $(U_t f)(p) = e^{itp} f(p)$. Then $(\theta f)(p) := f(-p)$ is a unitary involution with $\theta U_t \theta^{-1} = U_{-t}$ and $(Jf)(p) := \overline{f(-p)}$ is an anti-unitary involution with $J U_t J^{-1} = U_t$ for $t \in \mathbb{R}$. Clearly, θ and J commute.

For a systematic discussion of anti-unitary representations we refer to [NÓ17].

Theorem 5.1.7 (KMS Characterization Theorem; [NÓ16, Theorem 2.6]) *Let V be a real vector space, let $\beta > 0$, and let $\psi : \mathbb{R} \to \mathrm{Bil}(V)$ be a pointwise continuous positive definite function. Then the following are equivalent:*

  (i) *ψ satisfies the β-KMS condition.*
  (ii) *There exists a standard subspace V_1 in a Hilbert space $\mathcal{H}$ and a linear map $j : V \to V_1$ such that*

$$\psi(t)(\xi, \eta) = \langle j(\xi), \Delta_V^{-it/\beta} j(\eta) \rangle \quad \text{for} \quad t \in \mathbb{R}, \xi, \eta \in V. \qquad (5.7)$$

  (iii) *There exists a (uniquely determined) regular Borel measure μ on $\mathbb{R}$ with values in the cone $\mathrm{Bil}^+(V) \subseteq \mathrm{Bil}(V)$, consisting of forms with a positive semidefinite extension to $V_\mathbb{C}$, which satisfies $d\mu(-\lambda) = e^{-\beta\lambda} d\overline{\mu}(\lambda)$ and*

$$\psi(t) = \int_\mathbb{R} e^{it\lambda} d\mu(\lambda) \quad \text{for} \quad t \in \mathbb{R}.$$

If these conditions are satisfied, then the function $\psi : \overline{\mathscr{S}_\beta} \to \mathrm{Bil}(V)$ *is pointwise bounded.*

The equivalence of (i) and (ii) in this theorem describes the tight connection between the KMS condition and the modular objects associated to a standard subspace. Part (iii) provides an integral representation that can be viewed as a classification result in the sense that it characterizes those measures whose Fourier transforms satisfy the KMS condition from the perspective of Bochner's Theorem (Theorem A.2.1).

Example 5.1.8 If $V = \mathbb{R}$ and $\mathrm{Bil}(V) \cong \mathbb{C}$, $\mathrm{Bil}^+(V) = \mathbb{R}_{\geq 0}$, the integral representation in Theorem 5.1.7(iii) specializes to the integral representation obtained in Remark 4.1.7 for β-periodic reflection positive functions on $\mathbb{R}$:

$$\varphi(t) := \psi(it) = \mathscr{L}(\mu)(t) \quad \text{for} \quad 0 \leq t \leq \beta,$$

for a finite measure μ on $\mathbb{R}$ that can be written as $d\mu(\lambda) = d\mu_+(\lambda) + e^{\beta\lambda}d\mu_+(-\lambda)$ for a measure μ_+ on $\mathbb{R}_{\geq 0}$. This shows already that, in this case, the β-periodic extension of the function φ to $\mathbb{R}$ is reflection positive. Below we shall see how this observation can be extended to the general case.

The corresponding Hilbert space can be identified with $\mathscr{H} = L^2(\mathbb{R}, \mu)$, where $(U_t f)(\lambda) = e^{it\lambda} f(\lambda)$, so that $U_t = \Delta^{-it/\beta}$ leads to the modular operator

$$(\Delta f)(\lambda) = e^{-\beta\lambda} f(\lambda).$$

As μ is finite, $1 \in \mathscr{H}$ and we have $\psi(t) = \langle 1, U_t 1 \rangle$ for $t \in \mathbb{R}$. To determine a suitable standard subspace V_1, respectively, a conjugation J commuting with U, we note that the requirement $1 \in V_1$ and the requirement that J commutes with U lead to

$$(Jf)(\lambda) = e^{-\lambda\beta/2}\overline{f(-\lambda)},$$

so that the corresponding operator $S := J\Delta^{1/2}$ is given by $(Sf)(\lambda) = \overline{f(-\lambda)}$, and this leads to

$$V_1 = \{f \in L^2(\mathbb{R}, \mu) : f(-\lambda) = \overline{f(\lambda)} \ \mu\text{-almost everywhere}\}.$$

We shall continue the discussion of this example in Remark 5.2.6 below.

Remark 5.1.9 The KMS condition is well known in Quantum Statistical Mechanics as a condition characterizing quantum versions of Gibbs states, resp., equilibrium states. The monograph [BR96] and the lecture notes [Fro11] are excellent sources for more information on KMS states and their applications.

We now explain how the classical context of KMS states of operator algebras relates to our setup. Consider a C^*-*dynamical system* $(\mathscr{A}, \mathbb{R}, \alpha)$, i.e., a homomorphism $\alpha : \mathbb{R} \to \mathrm{Aut}(\mathscr{A})$, where $\mathscr{A}$ is a C^*-algebra. Here we deal with the real linear space

$$V := \mathscr{A}_h := \{A \in \mathscr{A} : A^* = A\}$$

of hermitian elements in $\mathscr{A}$, so that any state $\omega \in \mathscr{A}^*$ defines an element of $\mathrm{Bil}(V)$ by $(A, B) \mapsto \omega(AB)$. An α-invariant state ω on $\mathscr{A}$ is called a β-*KMS state* if and only if

$$\psi: \mathbb{R} \to \mathrm{Bil}(\mathscr{A}_h), \quad \psi(t)(A, B) := \omega(A\alpha_t(B))$$

satisfies the β-KMS condition (cf. [NÓ15b, Proposition 5.2], [RvD77, Theorem 4.10]). If $(\pi_\omega, U^\omega, \mathscr{H}_\omega, \Omega)$ is the corresponding covariant GNS representation of $(\mathscr{A}, \mathbb{R}, \alpha)$ (cf. [BGN17, BR02]), then

$$\omega(A) = \langle \Omega, \pi_\omega(A)\Omega \rangle \quad \text{for} \quad A \in \mathscr{A} \quad \text{and} \quad U^\omega_t \Omega = \Omega \quad \text{for} \quad t \in \mathbb{R}.$$

Therefore

$$\psi(t)(A, B) = \omega(A\alpha_t(B)) = \langle \Omega, \pi_\omega(A\alpha_t(B))\Omega \rangle$$
$$= \langle \Omega, \pi_\omega(A)U^\omega_t \pi_\omega(B)U^\omega_{-t}\Omega \rangle = \langle \pi_\omega(A)\Omega, U^\omega_t \pi_\omega(B)\Omega \rangle$$

for $A, B \in \mathscr{A}_h$. We conclude that the corresponding standard subspace of $\mathscr{H}_\omega$ is $V_1 := \pi_\omega(\mathscr{A}_h)\Omega$.

Corollary 5.1.10 *If $\psi: \mathbb{R} \to \mathrm{Bil}(V)$ satisfies the β-KMS condition, then the kernel*

$$K: \overline{\mathscr{S}_{\beta/2}} \times \overline{\mathscr{S}_{\beta/2}} \to \mathrm{Bil}(V), \qquad K(z, w)(\xi, \eta) := \psi(z - \overline{w})(\xi, \eta) \qquad (5.8)$$

is positive definite.

Proof From (5.7) in the KMS Characterization Theorem 5.1.7, we obtain by uniqueness of analytic continuation

$$\psi(z - \overline{w})(\xi, \eta) = \langle \Delta^{i\overline{z}/\beta} j(\xi), \Delta^{i\overline{w}/\beta} j(\eta) \rangle, \qquad \xi, \eta \in V, \ z, w \in \overline{\mathscr{S}_{\beta/2}}. \quad (5.9)$$

Now Remark A.1.2 shows that K is positive definite. □

Now that we know from Corollary 5.1.10 that the kernel K in (5.8) is positive definite, we obtain a corresponding reproducing kernel Hilbert space consisting of functions on $\overline{\mathscr{S}_{\beta/2}} \times V$ which are linear in the second argument and holomorphic on $\mathscr{S}_{\beta/2}$ in the first. We may therefore think of these functions as having values in the algebraic dual space $V^* := \mathrm{Hom}(V, \mathbb{R})$ of V. We write $\mathscr{O}(\overline{\mathscr{S}_{\beta/2}}, V^*)$ for the space of functions $f: \overline{\mathscr{S}_{\beta/2}} \to V^*$ for which all function $f^\eta(z) := f(z)(\eta), \eta \in V$, are continuous on $\overline{\mathscr{S}_{\beta/2}}$ and holomorphic on the open strip $\mathscr{S}_{\beta/2}$. For a proof of the following proposition, see [NÓ16, Proposition 2.9].

Proposition 5.1.11 (Holomorphic realization of $\mathcal{H}_\psi$) *Assume that* $\psi : \mathbb{R} \to \mathrm{Bil}(V)$ *satisfies the β-KMS condition, let* $\psi : \overline{\mathcal{S}_\beta} \to \mathrm{Bil}(V)$ *denote the corresponding extension and* $\mathcal{H}_\psi \subseteq \mathcal{O}(\overline{\mathcal{S}_{\beta/2}}, V^*)$ *denote the Hilbert space with reproducing kernel*

$$K(z, w)(\xi, \eta) := \psi(z - \overline{w})(\xi, \eta) \quad \text{for} \quad \xi, \eta \in V,$$

i.e.,

$$f(z)(\xi) = \langle K_{z,\xi}, f \rangle \quad \text{for} \quad f \in \mathcal{H}_\psi, \quad \text{where} \quad K_{z,\xi}(w)(\eta) = \psi(w - \overline{z})(\eta, \xi).$$

Then

$$(U_t^\psi f)(z) := f(z + t), \qquad t \in \mathbb{R}, z \in \overline{\mathcal{S}_{\beta/2}}$$

defines a unitary one-parameter group on $\mathcal{H}_\psi$,

$$j : V \to \mathcal{H}_\psi, \quad j(\eta)(z) := \psi(z)(\cdot, \eta) = K_{0,\eta}(z)$$

is a linear map with U^ψ-*cyclic range, and*

$$\psi(t)(\xi, \eta) = \langle j(\xi), U_t^\psi j(\eta) \rangle \quad \text{for} \quad t \in \mathbb{R}, \xi, \eta \in V.$$

The anti-unitary involution J_1 *on* $\mathcal{H}_\psi$ *corresponding to the standard subspace* $V_1 \subseteq \mathcal{H}_\psi$ *from* Theorem 5.1.7 *is given by* $(J_1 f)(z) := \overline{f\left(\overline{z} + \frac{i\beta}{2}\right)}$.

5.2 Reflection Positive Functions and KMS Conditions

In this section we build the bridge from positive definite functions $\psi : \mathbb{R} \to \mathrm{Bil}(V)$ satisfying the β-KMS condition to reflection positive functions on the group $\mathbb{T}_{2\beta,\tau} \cong O_2(\mathbb{R})$.

We have already seen in Lemma 5.1.2 that analytic continuation leads to a symmetric 2β-periodic function $\varphi : \mathbb{R} \to \mathrm{Bil}(V)$ satisfying $\varphi(t + \beta) = \overline{\varphi(t)}$ for $t \in \mathbb{R}$ and $\varphi(t) = \psi(it)$ for $0 \le t \le \beta$. We shall construct a positive definite extension $f : \mathbb{R}_\tau \to \mathrm{Bil}(V)$ with $f(t, \tau) = \varphi(t)$ for $t \in \mathbb{R}$; actually the values of f will be represented by bounded operators on $V_\mathbb{C}$, so that we also consider it as a $B(V_\mathbb{C})$-valued function. By construction, f is then reflection positive with respect to the interval $[0, \beta/2] =: G_+ \subseteq G := \mathbb{R}$ in the sense of Definition 3.4.1.

Building on Theorem 5.1.7, our first goal is to express, for a standard subspace $V \subseteq \mathcal{H}$, the $\mathrm{Bil}(V)$-valued function

$$\varphi_V : [0, \beta] \to \mathrm{Bil}(V), \qquad \varphi_V(t)(\xi, \eta) := \psi(it)(\xi, \eta) = \langle \Delta^{t/2\beta} \xi, \Delta^{t/2\beta} \eta \rangle \quad (5.10)$$

from (5.9) in terms of a $B(V_{\mathbb{C}})$-valued function. To this end, we shall need the description of a standard subspace V_1 in terms of a skew-symmetric *strict contraction* C on V_1 ($\|Cv\| < \|v\|$ for $0 \neq v$), and this leads to a quite explicit description of φ that is used to obtain the main theorem asserting that, for every positive definite function $\psi \colon \mathbb{R} \to \mathrm{Bil}(V)$ satisfying the β-KMS condition, there exists a reflection positive function $f \colon \mathbb{R}_\tau \to B(V_{\mathbb{C}})$ satisfying

$$\psi(it)(\xi, \eta) = \langle \xi, f(it, \tau)\eta \rangle \quad \text{for} \quad \xi, \eta \in V, 0 \leq t \leq \beta.$$

Then the corresponding GNS representation $(U^f, \mathcal{H}_f)$ of the group $\mathbb{T}_{2\beta,\tau} \cong O_2(\mathbb{R})$ is a euclidean realization of the unitary one-parameter group $(\Delta^{-it/\beta})_{t \in \mathbb{R}}$ corresponding to ψ via (5.7) because $\mathcal{E} = \mathcal{H}_f$ leads to $\widehat{\mathcal{E}} \cong \mathcal{H}_{\varphi|_{(0,\beta)}} \cong \mathcal{H}_\psi$ (cf. Theorem 3.4.5).

The following lemma describes the complex-valued scalar product on a standard real subspace in terms of the corresponding modular objects (Δ, J). For $v, w \in V$, we write $\langle v, w \rangle_V := \mathrm{Re}\langle v, w \rangle_{\mathcal{H}}$.

Lemma 5.2.1 *Let $V \subseteq \mathcal{H}$ be a standard subspace. Then there exists a skew-symmetric strict contraction C on V with*

$$\mathrm{Im}\langle \xi, \eta \rangle_{\mathcal{H}} = \langle \xi, C\eta \rangle_V \quad \text{for} \quad \xi, \eta \in V. \tag{5.11}$$

Proof Since $\omega(v, w) := \mathrm{Im}\langle v, w \rangle_{\mathcal{H}}$ defines a continuous skew-symmetric bilinear form on V, there exists a uniquely determined skew-symmetric operator $C \in B(V)$ with $\omega(v, w) = \langle v, Cw \rangle_V$ for $v, w \in V$. As $|\mathrm{Im}\langle v, w \rangle_{\mathcal{H}}| \leq \|v\| \cdot \|w\|$ for $v, w \in V$, we have $\|C\| \leq 1$, i.e., C is a contraction.

To see that C is a strict contraction, assume $\|Cv\| = \|v\|$, i.e., $v \in \ker(C^2 + 1)$. For $w := Cv$ we then have $C(v + iw) = w - iv = (-i)(v + iw)$. This leads to the relation $\langle v - iw, v - iw \rangle_{\mathcal{H}} = 0$ and thus $v - iw = 0$ implies $v \in V \cap iV = \{0\}$. $\square$

With the preceding lemma, we can express the function φ_V from (5.10) in terms of C by bounded operators on $V_{\mathbb{C}}$.

Lemma 5.2.2 ([NÓ16, Lemma 4.2]) *Let $V \subseteq \mathcal{H}$ be a standard subspace with modular objects (Δ, J) and C be the skew-symmetric strict contraction from Lemma 5.2.1. Then the function $\varphi_V(t)(\xi, \eta) = \langle \Delta^{t/2\beta}\xi, \Delta^{t/2\beta}\eta \rangle_{\mathcal{H}}$ from (5.10) can be written as*

$$\varphi_V(t)(\xi, \eta) = \langle \xi, \tilde{\varphi}(t)\eta \rangle_{V_{\mathbb{C}}} \quad \text{for} \quad t \in [0, \beta], \xi, \eta \in V_{\mathbb{C}} \tag{5.12}$$

with

$$\tilde{\varphi}(t) = (1 + iC)^{1-t/\beta}(1 - iC)^{t/\beta} \in B(V_{\mathbb{C}}).$$

Note that $\tilde{\varphi}(0) = 1 + iC$ is not real if $C \neq 0$ and that both operators $1 \pm iC$ are bounded positive hermitian with a possibly unbounded inverse. Therefore

$$\tilde{\psi}(z) = (\mathbf{1} + iC)^{1 + iz/\beta}(\mathbf{1} - iC)^{-iz/\beta} \in B(V_{\mathbb{C}})$$

is well-defined for $0 \leq \operatorname{Im} z \leq \beta$, strongly continuous and holomorphic for $0 < \operatorname{Im} z < \beta$. One also verifies immediately the β-KMS relation

$$\overline{\tilde{\psi}(z)} = \tilde{\psi}(i\beta + \overline{z}) \quad \text{for} \quad 0 \leq \operatorname{Re} z \leq \beta.$$

Theorem 5.2.3 (Reflection positive extension) *Let $V \subseteq \mathscr{H}$ be a standard subspace and let $C = I|C| \in B(V)$ be the skew-symmetric strict contraction satisfying (5.11). We assume that $\ker C = \{0\}$, so that I defines a complex structure on V. We define a weakly continuous function $\tilde{\varphi} \colon \mathbb{R} \to B(V_{\mathbb{C}})$ by*

$$\tilde{\varphi}(t) = (\mathbf{1} + iC)^{1 - t/\beta}(\mathbf{1} - iC)^{t/\beta} \quad \text{for} \quad 0 \leq t \leq \beta \quad \text{and} \quad \tilde{\varphi}(t + \beta) = \overline{\tilde{\varphi}(t)}$$

for $t \in \mathbb{R}$. Write $\tilde{\varphi}(t) = u^+(t) + iIu^-(t)$ with $u^\pm(t) \in B(V)$ and $u^\pm(t + \beta) = \pm u^\pm(t)$. Then

$$f \colon \mathbb{R}_\tau \to B(V_{\mathbb{C}}), \qquad f(t, \tau^\varepsilon) := u^+(t) + (iI)^\varepsilon u^-(t), \qquad t \in \mathbb{R}, \varepsilon \in \{0, 1\},$$

is a weak-operator continuous positive definite function with $f(t, \tau) = \tilde{\varphi}(t)$ for $t \in \mathbb{R}$. It is reflection positive with respect to the subset $[0, \beta/2] \subseteq \mathbb{R}$ in the sense that the kernel $f\big((t, \tau)(-s, e)\big) = f(t + s, \tau), 0 \leq s, t \leq \beta/2$, is positive definite.

Combining the preceding theorem with Lemma 5.2.2, we obtain in particular:

Corollary 5.2.4 *Let V be a real vector space and let $\psi \colon \mathbb{R} \to \operatorname{Bil}(V)$ be a continuous positive definite function satisfying the β-KMS condition. Then there exists a pointwise continuous function $f \colon \mathbb{R}_\tau \to \operatorname{Bil}(V)$ which is reflection positive with respect to the subset $[0, \beta/2] \subseteq \mathbb{R}$ and satisfies*

$$f(t, \tau) = \psi(it) \quad \text{for} \quad 0 \leq t \leq \beta \quad \text{and} \quad f(t + \beta, \tau) = \overline{f(t, \tau)} \quad \text{for} \quad t \in \mathbb{R}.$$

In Theorem 5.2.3 we obtained for certain functions φ on the coset $\mathbb{R} \rtimes \{\tau\} \subseteq \mathbb{R}_\tau$ reflection positive extensions f to all of $\mathbb{R}_\tau$. The following lemma shows that, conversely, every reflection positive function on $\mathbb{T}_{2\beta, \tau}$ leads by analytic extension to a positive definite function on $\mathbb{R}$ satisfying the β-KMS condition.

Lemma 5.2.5 *Let $f \colon \mathbb{R}_\tau \to \operatorname{Bil}(V)$ be a pointwise continuous function which is reflection positive with respect to $[0, \beta/2] \subseteq \mathbb{R}$ such that the function $\varphi \colon \mathbb{R} \to \operatorname{Bil}(V), \varphi(t) := f(t, \tau)$ satisfies*

$$\varphi(t) = \varphi(-t) = \overline{\varphi(\beta + t)} \quad \text{for} \quad t \in \mathbb{R}. \tag{5.13}$$

Then there exists a unique β-KMS positive definite function $\psi \colon \mathbb{R} \to \operatorname{Bil}(V)$ with

$$\varphi(t) = \psi(it) \quad \text{for} \quad 0 \leq t \leq \beta.$$

Proof Reflection positivity implies that the kernel $\varphi\left(\frac{t+s}{2}\right)$ for $0 \le t, s \le \beta$ is positive definite. By Theorem A.2.3 there exists a $\mathrm{Bil}^+(V)$-valued Borel measure μ on $\mathbb{R}$ such that

$$\varphi(t) = \int_{\mathbb{R}} e^{-\lambda t}\, d\mu(\lambda) \quad \text{for} \quad 0 < t < \beta. \tag{5.14}$$

The continuity of φ on $[0, \beta]$ actually implies that the integral representation also holds on the closed interval $[0, \beta]$ by the Monotone Convergence Theorem. In particular, the measure μ is finite. Therefore its Fourier transform $\psi(t) := \int_{\mathbb{R}} e^{it\lambda}\, d\mu(\lambda)$ is a pointwise continuous $\mathrm{Bil}(V)$-valued positive definite function on $\mathbb{R}$. Further, (5.13) implies

$$e^{\beta\lambda}\, d\mu(-\lambda) = d\overline{\mu}(\lambda) \tag{5.15}$$

and Theorem 5.2.3 shows that $\varphi(t) = \psi(it)$ holds for the β-KMS function $\psi \colon \mathbb{R} \to \mathrm{Bil}(V)$. $\qquad\qquad\square$

Remark 5.2.6 From (5.14) it follows that the function φ is real-valued if and only if the measure μ takes values in the subspace of real-valued forms in $\mathrm{Bil}^+(V)$.

For the case where $V \subseteq \mathscr{H}$ is a standard subspace and $\varphi = \varphi_V$ as in (5.10), we have $\tilde{\varphi}_V(0) = \mathbf{1} + iC$, so that $C = 0$ if φ_V is real-valued, and this in turn implies that φ_V is constant.

Therefore the only way to obtain non-constant real-valued functions is to ensure that the map $j \colon V \to V_1$ in Theorem 5.1.7 takes values in a subspace $j(V)$ which is isotropic for the skew-symmetric form $\omega(\xi, \eta) := \langle \xi, C\eta\rangle_V = \mathrm{Im}\langle\xi, \eta\rangle_{\mathscr{H}}$. This condition corresponds to $\varphi(0)$ being real, but is still weaker than $\varphi(t)$ being real for every $t \in [0, \beta]$.

If φ is real-valued, then $f(t, \tau^\varepsilon) := \tilde{\varphi}(t)$ for $t \in \mathbb{R}$, $\varepsilon \in \{0, 1\}$ is τ-biinvariant, β-periodic and reflection positive on $\mathbb{R}_\tau$ (Lemma 3.4.3(ii)).

It is instructive to take another look at Example 5.1.8, where $\mathscr{H} = L^2(\mathbb{R}, \mu)$ for a finite measure satisfying $d\mu(\lambda) = d\mu_+(\lambda) + e^{\beta\lambda} d\mu_+(-\lambda)$ for a measure μ_+ on $\mathbb{R}_{\ge 0}$. Here the standard subspace V_1 consists of all functions satisfying $f(-\lambda) = \overline{f(\lambda)}$ almost everywhere on $\mathbb{R}$. For simplicity we assume that $\mu(\{0\}) = 0$ (which excludes constant summands). For $\xi \in V_1$, the restriction $\xi_+ := \xi|_{\mathbb{R}_+}$ determines ξ completely, so that we may consider V_1 as a space of functions on $\mathbb{R}_+$. The scalar product on this space is given by

$$\langle \xi, \eta\rangle_{\mathscr{H}} = \int_{\mathbb{R}} \overline{\xi(\lambda)}\eta(\lambda)\, d\mu(\lambda) = \int_0^\infty \left(\overline{\xi_+(\lambda)}\eta_+(\lambda) + \xi_+(\lambda)\overline{\eta_+(\lambda)}e^{-\beta\lambda}\right) d\mu_+(\lambda).$$

For the real part we obtain

$$\langle \xi, \eta\rangle_{V_1} = \mathrm{Re}\langle\xi, \eta\rangle_{\mathscr{H}} = \int_0^\infty \mathrm{Re}\left(\overline{\xi_+(\lambda)}\eta_+(\lambda)\right)(1 + e^{-\beta\lambda})\, d\mu_+(\lambda),$$

and

$$\omega(\xi, \eta) = \operatorname{Im}\langle \xi, \eta \rangle_{\mathscr{H}} = \int_0^\infty \operatorname{Im}\left(\overline{\xi_+(\lambda)}\eta_+(\lambda)\right)(1 - e^{-\beta\lambda}) \, d\mu_+(\lambda)$$

for the imaginary part. We conclude that

$$V_1 \cong L^2(\mathbb{R}_+, (1 + e^{-\beta\lambda}) \, d\mu_+(\lambda); \mathbb{C})$$

and that the skew-symmetric operator C representing ω is given by

$$(Cf)(\lambda) = C(\lambda)f(\lambda), \quad \text{where} \quad C(\lambda) = -i\frac{1 - e^{-\beta\lambda}}{1 + e^{-\beta\lambda}}$$

(cf. [NÓ16, Lemma B.9]). Hence the corresponding complex structure is given by $(If)(\lambda) = -if(\lambda)$ and $|C| = iC$ corresponds to multiplication with the positive function $iC(\lambda) = \frac{1-e^{-\beta\lambda}}{1+e^{-\beta\lambda}}$ on $\mathbb{R}_+$.

The subspace

$$V := L^2(\mathbb{R}_+, (1 + e^{-\beta\lambda}) \, d\mu_+(\lambda); \mathbb{R})$$

of real-valued functions is ω-isotropic. As it is invariant under the operators

$$\tilde{\varphi}(t) = (\mathbf{1} + iC)^{1-t/\beta}(\mathbf{1} - iC)^{t/\beta} = (\mathbf{1} + |C|)^{1-t/\beta}(\mathbf{1} - |C|)^{t/\beta},$$

the corresponding function

$$\varphi \colon [0, \beta] \to \operatorname{Bil}(V), \quad \varphi(t)(\xi, \eta) = \langle \xi, \tilde{\varphi}(t)\eta \rangle, \quad \xi, \eta \in V$$

is real-valued.

From the scalar case ($V = \mathbb{R}$) in Remark 4.1.7 one easily obtains the following characterization of β-periodic operator-valued reflection positive functions on $\mathbb{R}$. It is concerned with the case where φ is real-valued, so that f is τ-biinvariant (Lemma 3.4.3), corresponding to function on the circle group $\mathbb{T}_\beta$ (see also [KL81, Theorem 3.3]).

Theorem 5.2.7 *A β-periodic pointwise continuous function $\varphi \colon \mathbb{R} \to \operatorname{Bil}(V)$ is reflection positive with respect to $[0, \beta/2]$ if and only if there exists a $\operatorname{Bil}^+(V)$-valued Borel measure μ_+ on $[0, \infty)$ such that*

$$\varphi(t) = \int_0^\infty e^{-t\lambda} + e^{-(\beta-t)\lambda} \, d\mu_+(\lambda) \quad \text{for} \quad 0 \leq t \leq \beta. \tag{5.16}$$

Then the measure μ_+ is uniquely determined by φ.

Definition 5.2.8 (Euclidean realization in the periodic case) For any reflection positive function f as in Lemma 5.2.5, the general discussion in Theorem 3.4.5 shows that, for the corresponding reflection positive representation on $\mathscr{E} = \mathscr{H}_f$, we obtain $\widehat{\mathscr{E}} \cong \mathscr{H}_{\varphi|_{(0,\beta)}}$.

As ψ is pointwise holomorphic on the strip $\mathscr{S}_\beta$, it further follows by restriction that $\mathscr{H}_{\varphi|_{(0,\beta)}} \cong \mathscr{H}_\psi$ (cf. Proposition 5.1.11). Therefore the unitary one-parameter group $(U_t^c f)(z) := f(z + t)$ on $\widehat{\mathscr{E}}$ whose infinitesimal generator is given by $\frac{d}{dz}$, is obtained from the unitary representation U^f on $\mathscr{E}$ by the OS transform as in Example 3.3.5, even if it is not positive. We thus call $(U^f, \mathscr{H}_f)$ a *euclidean realization* of U^c (cf. Definition 3.3.4).

At this point it is a natural question which unitary one-parameter groups $(U^c, \mathscr{H})$ have a euclidean realization. This can now be stated in terms of the conditions discussed in Proposition 5.1.5 ([NÓ15b, Theorem 3.4]):

Theorem 5.2.9 (Realization Theorem) *A unitary one-parameter group $(U_t^c)_{t\in\mathbb{R}}$ on a Hilbert space $\mathscr{H}$ has a euclidean realization in terms of a reflection positive representation of $(\mathbb{T}_{2\beta}, \mathbb{T}_{2\beta,+}, \theta)$ if and only if there exists an anti-unitary involution J on $\mathscr{H}$ commuting with U^c.*

In the setting of Theorem 5.2.9, a particular euclidean realization can be obtained as follows. Let $U_t^c = e^{itH}$ be a unitary one-parameter group on $\mathscr{H}$ and J be a unitary involution on $\mathscr{H}$ with $JHJ = -H$. Then $\Delta := e^{-\beta H}$ satisfies $J\Delta J = \Delta^{-1}$, so that $V := \mathrm{Fix}(J\Delta^{1/2})$ is a standard subspace and Theorem 5.1.7 leads to a positive definite function $\psi\colon \mathbb{R} \to \mathrm{Bil}(V)$ satisfying the β-KMS condition. Now Theorem 5.2.3 yields a reflection positive function on $\mathbb{R}_\tau$, resp., $\mathbb{T}_{2\beta,\tau}$, which provides a euclidean realization of U^c.

5.3 Realization by Resolvents of the Laplacian

Before we describe a realization of the GNS representation $(U^f, \mathscr{H}_f)$ in spaces of sections of a vector bundle, let us recall the general background for this.

Remark 5.3.1 For a $B(V)$-valued positive definite function $f\colon G \to B(V)$, the reproducing kernel Hilbert space $\mathscr{H}_f = \mathscr{H}_K$ with kernel $K(g, h) = \varphi(gh^{-1}) = K_g K_h^*$ is generated by the functions

$$K_{h,w} := K_h^* w \quad \text{with} \quad K_{h,w}(g) = K_g K_h^* w = K(g, h)w = \varphi(gh^{-1})w.$$

The group G acts on this space by right translations

$$(U_g s)(h) := s(hg).$$

If $P \subseteq G$ is a subgroup and (ρ, V) is a unitary representation of P such that

$$f(hg) = \rho(h)f(g) \quad \text{for all} \quad g \in G, h \in P,$$

then

$$\mathscr{H}_f \subseteq \mathscr{F}(G, V)_\rho := \{s : G \to V : (\forall g \in G)(\forall h \in P) s(hg) = \rho(h)s(g)\}.$$

Therefore $\mathscr{H}_f$ can be identified with a space of sections of the associated vector bundle

$$\mathbb{V} := (V \times_P G) = (V \times G)/P,$$

where P acts on the trivial vector bundle $V \times G$ over G by $h.(v, g) = (\rho(h)v, hg)$.

To derive a suitable characterization of the functions f arising in Theorem 5.2.3, we identify 2β-periodic functions s on $\mathbb{R}$ via $s = s_+ + s_-$ with pairs of functions (s_+, s_-) satisfying $s_\pm(\beta + t) = \pm s_\pm(t)$. Accordingly, any 2β-periodic function $s : \mathbb{R} \to V_\mathbb{C}$ defines a function

$$\tilde{s} : \mathbb{R} \to V_\mathbb{C}^2, \quad \tilde{s} = (s_+, s_-) \quad \text{with} \quad \tilde{s}(\beta + t) = \begin{pmatrix} 1 & 0 \\ 0 & -1 \end{pmatrix} \tilde{s}(t).$$

In this sense $\tilde{s}$ is a section of the vector bundle over $\mathbb{T}_\beta$ with fiber $V_\mathbb{C}^2$ defined by the representation of $\beta\mathbb{Z}$, specified by $\rho(\beta) = \begin{pmatrix} 1 & 0 \\ 0 & -1 \end{pmatrix}$. Splitting the $B(V_\mathbb{C})$-valued positive definite function

$$f : \mathbb{R}_\tau \to B(V_\mathbb{C}), \quad f(t, \tau^\varepsilon) = u^+(t) + u^-(t)(iI)^\varepsilon \quad \text{for} \quad t \in \mathbb{R}, \varepsilon \in \{0, 1\}$$

as in Theorem 5.2.3 into even and odd part with respect to the β-translation, we obtain the following lemma which shows in particular that we may identify the Hilbert space $\mathscr{H}_f \cong \mathscr{H}_{f^\sharp}$ as a space of section of a Hilbert bundle $V_\mathbb{C}^2 \times_\rho \mathbb{R}_\tau$ over the circle $\mathbb{T}_\beta$ with fiber V^2.

Lemma 5.3.2 *For the subgroup* $P := (\mathbb{Z}\beta)_\tau \cong \mathbb{Z}\beta \rtimes \{e, \tau\}$ *of* $G := \mathbb{R}_\tau$, *we consider the unitary representation* $\rho : P \to U(V_\mathbb{C}^2)$ *defined by*

$$\rho(\beta, e) := \begin{pmatrix} 1 & 0 \\ 0 & -1 \end{pmatrix} \quad \text{and} \quad \rho(0, \tau) := \begin{pmatrix} 1 & 0 \\ 0 & iI \end{pmatrix},$$

where I *is the complex structure from the polar decomposition* $C = I|C|$ *on the real Hilbert space* V. *Then*

$$f^\sharp : \mathbb{R}_\tau \to B(V_\mathbb{C}^2) \cong M_2(B(V_\mathbb{C})), \quad f^\sharp(t, \tau^\varepsilon) := \begin{pmatrix} u^+(t) & 0 \\ 0 & u^-(t)(iI)^\varepsilon \end{pmatrix}$$

is a positive definite function satisfying

$$f^\sharp(hg) = \rho(h)f^\sharp(g) \quad \text{for} \quad h \in P, g \in G. \tag{5.17}$$

The corresponding GNS representation $(U^{f^\sharp}, \mathcal{H}_{f^\sharp})$ is equivalent to the GNS representation $(U^f, \mathcal{H}_f)$.

Proof The first assertion follows from

$$f^\sharp((0,\tau)(t,\tau^\varepsilon)) = \begin{pmatrix} u^+(-t) & 0 \\ 0 & u^-(-t)(iI)^{\varepsilon+1} \end{pmatrix} = \begin{pmatrix} u^+(t) & 0 \\ 0 & u^-(t)(iI)^{\varepsilon+1} \end{pmatrix}$$

and

$$f^\sharp(\beta+t,\tau^\varepsilon) = \begin{pmatrix} u^+(t) & 0 \\ 0 & -u^-(t)(iI)^\varepsilon \end{pmatrix}.$$

As the GNS representation $(U^f, \mathcal{H}_f)$ decomposes under the unitary involution U^f_β into the ± 1-eigenspaces, it is equivalent to the GNS representation $(U^{f^\sharp}, \mathcal{H}_{f^\sharp})$ corresponding to $f^\sharp$. $\qquad\square$

Remark 5.3.3 (a) In view of (5.15), there exists a $\mathrm{Bil}^+(V)$-valued measure ν on $[0,\infty)$ for which we can write $d\mu(\lambda) = d\nu(\lambda) + e^{\beta\lambda}d\overline{\nu}(-\lambda)$. For $\nu = \nu_1 + i\nu_2$, this leads for $0 \le t \le \beta$ to

$$\varphi(t) = \int_0^\infty e^{-t\lambda} + e^{-(\beta-t)\lambda}\,d\nu_1(\lambda) + i\int_0^\infty e^{-t\lambda} - e^{-(\beta-t)\lambda}\,d\nu_2(\lambda). \tag{5.18}$$

In particular, the most basic examples correspond to Dirac measures of the form $\nu = \delta_\lambda \cdot (\gamma + i\omega)$, where δ_λ is the Dirac measure in $\lambda > 0$:

$$\varphi(t) = (e^{-t\lambda} + e^{-(\beta-t)\lambda})\gamma + i(e^{-t\lambda} - e^{-(\beta-t)\lambda})\omega = e^{-t\lambda}h + e^{-(\beta-t)\lambda}\overline{h},$$

where $h := \gamma + i\omega \in \mathrm{Bil}^+(V)$.

Writing $\omega(\xi,\eta) = \gamma(\xi, C\eta)$ (Lemma 5.2.1) and replacing V by the real Hilbert space defined by the positive semidefinite form γ on V, we obtain the $B(V_{\mathbb{C}})$-valued function

$$\tilde{\varphi}(t) = (e^{-t\lambda} + e^{-(\beta-t)\lambda})\mathbf{1} + (e^{-t\lambda} - e^{-(\beta-t)\lambda})iC = e^{-t\lambda}(\mathbf{1}+iC) + e^{-(\beta-t)\lambda}(\mathbf{1}-iC)$$

for $0 \le t \le \beta$. This leads to

$$f(t,\tau^\varepsilon) = (1 + e^{-\beta\lambda})(u_\lambda^+(t)\mathbf{1} + u_\lambda^-(t)|C|(iI)^\varepsilon) \quad \text{for} \quad t \in \mathbb{R}, \varepsilon \in \{0,1\},$$

where

$$u_\lambda^\pm(t) = \frac{e^{-t\lambda} \pm e^{-(\beta-t)\lambda}}{1 + e^{-\beta\lambda}} \quad \text{for} \quad 0 \le t \le \beta, \qquad u_\lambda^\pm(t+\beta) = \pm u_\lambda^\pm(t).$$

(b) This can also be formulated in terms of forms. With $\gamma(\xi, \eta) = \langle \xi, \eta \rangle_V$ and

$$h(\xi, \eta) = \gamma(\xi, \eta) + i\omega(\xi, \eta) = \langle \xi, (\mathbf{1} + iC)\eta \rangle_{V_{\mathbb{C}}} = \langle \xi, (\mathbf{1} + iI|C|)\eta \rangle_{V_{\mathbb{C}}},$$

we get $f(t, \tau^\varepsilon)(\xi, \eta) = (1 + e^{-\beta\lambda}) \langle \xi, \left(u_\lambda^+(t)\mathbf{1} + u_\lambda^-(t)|C|(iI)^\varepsilon\right)\eta \rangle.$

We have seen above how to obtain a realization of the Hilbert space $\mathscr{H}_f$ as a space $\mathscr{H}_{f^\sharp}$ of sections of a Hilbert bundle $\mathbb{V}$ with fiber $V_{\mathbb{C}}^2$ over the circle $\mathbb{T}_\beta = \mathbb{R}/\beta\mathbb{Z}$. In this section we provide an analytic description of the scalar product on this space if $|C| = \mu\mathbf{1}, 0 < \mu < 1$, so that $\frac{1+|C|}{1-|C|} = e^\lambda\mathbf{1}$ for $\lambda := \log\left(\frac{1+\mu}{1-\mu}\right) > 0$. We shall see that it has a natural description in terms of the resolvent $(\lambda^2 - \Delta)^{-1}$ of the Laplacian Δ of $\mathbb{T}_\beta$ acting on section of the bundle $\mathbb{V}$.

As in Lemma 5.3.2, we write

$$f^\sharp(t, \tau^\varepsilon) = \begin{pmatrix} u_\lambda^+(t)\mathbf{1} & 0 \\ 0 & u_\lambda^-(t)(iI)^\varepsilon \end{pmatrix} \in B(V_{\mathbb{C}}^2) \cong M_2(B(V_{\mathbb{C}})),$$

For $\chi_n(t) = e^{\pi i n t/\beta}$ we then have $u_\lambda^+ = \sum_{n \in \mathbb{Z}} c_{2n}^\lambda \chi_{2n}$ and $u_\lambda^- = \sum_{n \in \mathbb{Z}} c_{2n+1}^\lambda \chi_{2n+1}$, where

$$c_n^\lambda = c_{-n}^\lambda = \frac{1 - (-1)^n e^{-\beta\lambda}}{1 + e^{-\beta\lambda}} \cdot \frac{2\beta\lambda}{(\beta\lambda)^2 + (n\pi)^2} = \frac{1 - (-1)^n e^{-\beta\lambda}}{1 + e^{-\beta\lambda}} \cdot \frac{2\lambda}{\beta} \cdot \frac{1}{\lambda^2 + (n\pi/\beta)^2}$$

for $n \in \mathbb{Z}$ (the rightmost factors are called bosonic Matsubara coefficients if n is even and fermionic if n is odd [DG13, Sect. 18]). With

$$c_+^\lambda := \frac{1 - e^{-\beta\lambda}}{1 + e^{-\beta\lambda}} \frac{2\lambda}{\beta} = \tanh\left(\frac{\beta\lambda}{2}\right)\frac{2\lambda}{\beta} \quad \text{and} \quad c_-^\lambda := \frac{2\lambda}{\beta}, \tag{5.19}$$

we thus obtain

$$c_{2n}^\lambda = \frac{c_+^\lambda}{\lambda^2 + (2n\pi/\beta)^2}, \quad c_{2n+1}^\lambda = \frac{c_-^\lambda}{\lambda^2 + ((2n+1)\pi/\beta)^2}. \tag{5.20}$$

The following proposition shows that the positive operator $(\lambda^2\mathbf{1} - \Delta_{\mathbb{R}})^{-1}$ on the Hilbert space of L^2-section of $\mathbb{V}$ defines a unitary representation of $\mathbb{R}_\tau$ which is unitarily equivalent to the representation on $\mathscr{H}_f$ (cf. Lemma 5.3.2).

Proposition 5.3.4 *For $\lambda > 0$, let $\mathscr{H}_\lambda$ be the Hilbert space obtained by completing the space*

$$\Gamma_\rho := \{s \in C^\infty(\mathbb{R}_\tau, V_{\mathbb{C}}^2) : (\forall g \in \mathbb{R}_\tau, h \in (\mathbb{Z}\beta)_\tau)\, s(hg) = \rho(h)s(g)\}$$

with respect to

$$\langle s_1, s_2 \rangle := \frac{1}{2\beta} \int_0^{2\beta} \langle s_1(t, e), ((\lambda^2 \mathbf{1} - \Delta_{\mathbb{R}})^{-1} s_2)(t, e) \rangle \, dt, \quad \text{where} \quad \Delta_{\mathbb{R}} = \frac{d^2}{dt^2}.$$

On $\mathscr{H}_\lambda$ we have a natural unitary representation U^λ of $\mathbb{R}_\tau$ by right translation which is unitarily equivalent to the GNS representation $(U^{f^\sharp}, \mathscr{H}_{f^\sharp})$. Here the corresponding inclusion map is given by

$$j \colon V \to \mathscr{H}_\lambda, \quad j \begin{pmatrix} v_1 \\ v_2 \end{pmatrix} = \sqrt{c_+^\lambda} \sum_{n \in \mathbb{Z}} \chi_{2n} \begin{pmatrix} v_1 \\ 0 \end{pmatrix} + \sqrt{c_-^\lambda} \sum_{n \in \mathbb{Z}} \chi_{2n+1} \begin{pmatrix} 0 \\ v_2 \end{pmatrix}. \tag{5.21}$$

This result provides a natural euclidean realization of our representation on the Riemannian manifold $\mathbb{T}_\beta \cong \mathbb{S}^1$ in the spirit of Theorem 2.5.1. For more recent work in this direction see [NÓ17, FNO18].

Remark 5.3.5 In the context of anti-unitary representations, it is interesting to observe that the reflection positive representation of $\mathbb{R}_\tau$, resp., $\mathbb{T}_{2\beta,\tau}$ described in Proposition 5.3.4 carries a natural anti-unitary involution given by

$$(Js)(t, \tau^\varepsilon) := \overline{s\left(\tfrac{\beta}{2} - t, \tau^\varepsilon\right)} \quad \text{for} \quad t \in \mathbb{R}, \varepsilon \in \{0, 1\}.$$

In fact, one readily verifies that J defines an anti-unitary involution on $\mathscr{H}_\lambda$. We further have $J U_\tau J = U_\tau$ and $J U_t J = U_{-t}$ for $t \in \mathbb{R}$.

Notes

The material in this chapter mainly draws from [NÓ16] which continued the investigations from [NÓ15b] only dealing with real-valued functions φ. This was motivated by the work of Klein and Landau in [KL81]. A long term goal is to combine our representation theoretic approach to reflection positivity with KMS states of operator algebras and Borchers triples corresponding to modular inclusions [NÓ17, BLS11, Bo92, Lo08, Sch99].

We have seen that the unitary one-parameter groups $(U^c, \widehat{\mathscr{E}})$ arising from reflection positivity on $\mathbb{T}_{2\beta}$ always commute with an anti-unitary involution. It would be nice to incorporate anti-unitary operators such as conjugations and anti-conjugations more systematically into the whole setup of the OS transform on the level of representations. This requires a better understanding of the role of anti-unitary operators on the euclidean side. Some first steps to a more systematic understanding of anti-unitary representations have been undertaken in [NÓ17, Ne18], but this has not yet been connected to reflection positivity.

For KMS states of the CCR (canonical commutation relations), resp. the corresponding Weyl algebra, we refer to the two papers of B. S. Kay [Ka85, Ka85b], dealing with uniqueness of KMS states for a given one-parameter group of symmetries and the embedding of KMS representations into irreducible ones by a doubling procedure (see also [BR96] for a more direct but less conceptual approach to the uniqueness of KMS states).

Interesting references for the relation of the KMS condition with (quantum) statistical mechanics are [Fro11] and [BR96].

Chapter 6
Integration of Lie Algebra Representations

A central problem in the context of reflection positive representations of a symmetric Lie group (G, τ) on a reflection positive Hilbert space $(\mathscr{E}, \mathscr{E}_+, \theta)$ is to construct on the associated Hilbert space $\widehat{\mathscr{E}}$ a unitary representations of the 1-connected Lie group G^c with Lie algebra $\mathfrak{g}^c = \mathfrak{h} + i\mathfrak{q}$. As we have seen in Remark 3.3.9, the main point is to "integrate" a unitary representation of the Lie algebra $\mathfrak{g}^c$ on a pre-Hilbert space. In general this problem need not have a solution, but we shall see below that in the reflection positive contexts, where the Hilbert spaces are mostly constructed from G-invariant positive definite kernels or positive definite G-invariant distributions, there are natural assumptions that apply in all cases that we consider.

For any reflection positive representation of (G, τ), we immediately obtain a unitary representation of the subgroup $H = G_0^\tau$ on $\widehat{\mathscr{E}}$, so that we have to find a unitary representation on the one-parameter group $\exp_{G^c}(\mathbb{R}iy)$ for $y \in \mathfrak{q}$. Since we have already a symmetric operator $\widehat{\mathrm{d}U}(x)$ on a dense subspace of $\widehat{\mathscr{E}}$, the essential point is to show that it is essentially selfadjoint.

In Sect. 6.1 we introduce Fröhlich's Theorem which provides a criterion for the essential selfadjointness of a symmetric operator. In Sect. 6.2 we connect this tool with the geometric context, where we consider a pair (β, σ) of a homomorphism $\beta : \mathfrak{g} \to \mathscr{V}(M)$ to the Lie algebra of smooth vector fields on a manifold M which is compatible with a smooth H-action σ. For any smooth kernel K on M satisfying a suitable invariance condition with respect to (β, σ), a unitary representation of G^c on $\mathscr{H}_K$ exists (Theorem 6.2.3). In Sect. 6.3 we show that this result remains valid if we replace the kernel K by a positive definite distribution $K \in C^{-\infty}(M \times M)$ compatible with (β, σ) (Theorem 6.3.6). We finally explain in Sect. 6.4 how these results apply to reflection positive representations.

Throughout this section M denotes a smooth manifold modeled on a Banach space, if not stated otherwise, and $\mathscr{V}(M)$ denotes the Lie algebra of smooth vector fields on M.

© The Author(s) 2018
K.-H. Neeb and G. Ólafsson, *Reflection Positivity*, SpringerBriefs in Mathematical Physics, https://doi.org/10.1007/978-3-319-94755-6_6

6.1 A Geometric Version of Fröhlich's Selfadjointness Theorem

We start with Fröhlich's Theorem on unbounded symmetric semigroups as it is stated in [Fro80, Cor. 1.2] (see also [MN12]). Actually Fröhlich assumes that the Hilbert space $\mathscr{H}$ is separable, but this is not necessary. Replacing the assumption of weak measurability by weak continuity, all arguments in [Fro80] work for non-separable spaces as well.

Theorem 6.1.1 (Fröhlich's Selfadjointness Theorem) *Let H be a symmetric operator defined on the dense subspace $\mathscr{D}$ of the Hilbert space $\mathscr{H}$. Suppose that, for every $\xi \in \mathscr{D}$, there exists an $\varepsilon_\xi > 0$ and a differentiable curve $\varphi : (0, \varepsilon_\xi) \to \mathscr{D}$ satisfying*

$$\varphi'(t) = H\varphi(t) \quad and \quad \lim_{t \to 0} \varphi(t) = \xi.$$

Then the operator H is essentially selfadjoint and $\varphi(t) = e^{t\overline{H}}\xi$ in the sense of spectral calculus of selfadjoint operators.

For later applications, we explain how Fröhlich's Theorem applies to linear vector fields on locally convex spaces. Let V be a locally convex space and the kernel $K : V \times V \to \mathbb{C}$ be a continuous positive semidefinite hermitian form. Then the corresponding reproducing kernel space $\mathscr{H}_K$ can be identified with a linear subspace of the space $V^\sharp$ of anti-linear continuous functionals on V (cf. Sect. A.1). It is generated by the functionals $K_w(v) := K(v, w), w \in V$, satisfying

$$\langle K_v, K_w \rangle = K_w(v) = K(v, w).$$

So it can also be interpreted as the completion of V with respect to the hermitian form K.

The continuity of the kernel K implies that the linear map $V \to \mathscr{H}_K$, $v \mapsto K_v$ is continuous. For any continuous linear operator $L\, V \to V$, the formula

$$L^K : \mathscr{D}_L \to \mathscr{H}_K, \quad L^K \lambda := -\lambda \circ L, \quad \mathscr{D}_L := \{\lambda \in \mathscr{H}_K \subseteq V^\sharp : \lambda \circ L \in \mathscr{H}_K\}$$

defines an unbounded closed operator on $\mathscr{H}_K$. If there exists an operator $L^* : V \to V$ with

$$K(v, Lw) = K(L^*v, w) \quad \text{for} \quad v, w \in V,$$

then we have

$$L^K K_v = K_{-L^*v} \quad \text{for} \quad v \in V. \tag{6.1}$$

We can now obtain from Theorem 6.1.1 ([MNO15, Cor. 4.9]):

Corollary 6.1.2 *Let* $L : V \to V$ *be a continuous linear operator on the locally convex space* V *which is* K*-symmetric in the sense that* $K(Lv, w) = K(v, Lw)$ *for* $v, w \in V$. *Suppose that, for every* $v \in V$, *there exists a curve* $\gamma_v : [0, \varepsilon_v] \to V$ *starting in* v *and satisfying the differential equation*

$$\gamma_v'(t) = L\gamma_v(t).$$

Then the restriction $L^K|_{\mathscr{H}_K^0}$ *to the dense subspace* $\mathscr{H}_K^0 = \{K_v : v \in V\} \subseteq \mathscr{H}_K$ *is essentially selfadjoint with closure* $\overline{L^K}$. *For* $0 \le t \le \varepsilon_v$, *we have* $e^{-t\overline{L^K}} K_v = K_{\gamma_v(t)}$.

Now we turn to the nonlinear setting of smooth positive definite kernels on manifolds. Here symmetric operators are obtained from smooth vector fields.

Definition 6.1.3 Let $K \in C^\infty(M \times M, \mathbb{C})$ be a smooth positive definite kernel and $\mathscr{H}_K \subseteq C^\infty(M)$ be the corresponding reproducing kernel Hilbert space.

(a) For a smooth vector field $X \in \mathscr{V}(M)$, we write

$$\mathscr{L}_X : C^\infty(M) \to C^\infty(M), \quad (\mathscr{L}_X f)(m) := \mathrm{d}f(m)X(m)$$

for the *Lie derivative on smooth functions*. We thus obtain on the reproducing kernel space $\mathscr{H}_K$ the unbounded operator

$$\mathscr{L}_X^K := \mathscr{L}_X|_{\mathscr{D}_X} : \mathscr{D}_X \to \mathscr{H}_K, \quad \text{where} \quad \mathscr{D}_X := \{\varphi \in \mathscr{H}_K : \mathscr{L}_X\varphi \in \mathscr{H}_K\}. \tag{6.2}$$

(b) A vector field $X \in \mathscr{V}(M)$ is said to be *K-symmetric (K-skew-symmetric)* if

$$\mathscr{L}_X^1 K = \varepsilon \mathscr{L}_X^2 K \quad \text{for} \quad \varepsilon = 1, \quad \text{resp.}, \quad \varepsilon = -1.$$

Here the superscripts indicate whether the Lie derivative acts on the first or the second argument of K.

The following theorem can be obtained quite directly from Fröhlich's Theorem if the Hilbert space under consideration has a smooth positive definite kernel ([MNO15, Theorem 4.6]):

Theorem 6.1.4 (Geometric Fröhlich Theorem) *Let* M *be a smooth manifold and* K *be a smooth positive definite kernel. If* X *is a* K*-symmetric vector field on* M, *then* $\mathscr{L}_X|_{\mathscr{H}_K^0}$ *is an essentially selfadjoint operator on* $\mathscr{H}_K$ *whose closure coincides with the operator* $\mathscr{L}_X^K$. *For* $m \in M$ *and an integral curve* $\gamma_m : [0, \varepsilon_m] \to M$ *of* X *with* $\gamma_m(0) = m$, *we have* $e^{t\mathscr{L}_X^K} K_m = K_{\gamma_m(t)}$ *for* $0 \le t \le \varepsilon_m$.

6.2 Integrability for Reproducing Kernel Spaces

We now turn to actions of a symmetric Lie algebra $(\mathfrak{g}, \tau)$ on a smooth manifold M that are compatible with a smooth positive definite kernel K. Our first main result is Theorem 6.2.3 which provides a sufficient condition for the Lie algebra representation of the dual Lie algebra $\mathfrak{g}^c$ coming from an action of $\mathfrak{g}$ on $\mathscr{H}_K$ by Lie derivatives to integrate to a unitary representation of the corresponding simply connected Lie group G^c. Applying this result to open subsemigroups of Lie groups, we further obtain an interesting generalization of the Lüscher–Mack Theorem [LM75, HN93] for semigroups which no longer requires the existence of a polar decomposition.

Definition 6.2.1 Let $(\mathfrak{g}, \tau)$ be a symmetric Lie algebra, and let $\beta : \mathfrak{g} \to \mathscr{V}(M)$ be a homomorphism. A smooth positive definite kernel $K \in C^\infty(M \times M, \mathbb{C})$ is said to be *β-compatible* if the vector fields in $\beta(\mathfrak{h})$ are K-skew-symmetric and the vector fields in $\beta(\mathfrak{q})$ are K-symmetric.

Definition 6.2.2 Let H be a connected Lie group with Lie algebra $\mathfrak{h}$. A *smooth right action* of the pair $(\mathfrak{g}, H)$ on M is a pair (β, σ), where

(a) $\sigma : M \times H \to M, (m, h) \mapsto \sigma_h(m) = m.h$ is a smooth right action,
(b) $\beta : \mathfrak{g} \to \mathscr{V}(M)$ is a homomorphism of Lie algebras, and
(c) $\mathrm{d}\sigma(x) = \beta(x)$ for $x \in \mathfrak{h}$.

In the following K denotes a smooth β-compatible positive definite kernel on $M \times M$. For $x \in \mathfrak{g}$, we abbreviate $\mathscr{L}_x := \mathscr{L}^K_{\beta(x)}$ for the maximal restriction of the Lie derivatives to $\mathscr{D}_x := \mathscr{D}_{\beta(x)}$ from (6.2) and we extend this definition in a complex linear fashion to $\mathfrak{g}_\mathbb{C}$. We also consider the subspace

$$\mathscr{D} := \{\varphi \in \mathscr{H}_K : (\forall n \in \mathbb{N})(\forall x_1, \ldots, x_n \in \mathfrak{g})\, \mathscr{L}_{\beta(x_1)} \cdots \mathscr{L}_{\beta(x_n)}\varphi \in \mathscr{H}_K\}$$

on which

$$\alpha : \mathfrak{g}_\mathbb{C} \to \mathrm{End}(\mathscr{D}), \ x \mapsto \mathscr{L}_x|_{\mathscr{D}}$$

defines a Lie algebra representation such that $\mathfrak{g}^c$ acts by skew-symmetric operators. The following theorem ([MNO15, Theorem 5.12]) asserts the integrability of $\alpha|_{\mathfrak{g}^c}$. Besides the usual technicalities, a key point in its proof is to apply the Geometric Fröhlich Theorem 6.1.4 to the vector fields $\beta(y)$, $y \in \mathfrak{q}$.

Theorem 6.2.3 *Let K be a smooth positive definite kernel on the manifold M compatible with the smooth right action (β, σ) of $(\mathfrak{g}, H)$, where $\mathfrak{g} = \mathfrak{h} \oplus \mathfrak{q}$ is a symmetric Lie algebra and H is a connected Lie group with Lie algebra $\mathfrak{h}$. Let G^c be a simply connected Lie group with Lie algebra $\mathfrak{g}^c = \mathfrak{h} + i\mathfrak{q}$. Then there exists a unique continuous unitary representation $(U^c, \mathscr{H}_K)$ such that*

(i) $\overline{\mathrm{d}U^c(x)} = \mathscr{L}^K_x$ *for $x \in \mathfrak{h}$.*
(ii) $\overline{\mathrm{d}U^c(iy)} = i\mathscr{L}^K_y$ *for $y \in \mathfrak{q}$.*

Remark 6.2.4 Note that (i) implies that the restriction of U^c to the integral subgroup $\langle \exp \mathfrak{h} \rangle \subseteq G^c$ induces the same representation on the universal covering group $\tilde{H}$ of H as the unitary representation $(U^H_h f)(m) := f(m.h)$ of H on $\mathscr{H}_K$ because their derived representations coincide (cf. Chap. 7).

Example 6.2.5 Let (G, τ) be a symmetric Lie group with Lie algebra $\mathfrak{g} = \mathfrak{h} + \mathfrak{q}$ and let $H = G_0^\tau$ be the integral subgroup corresponding to the Lie subalgebra $\mathfrak{h} = \mathfrak{g}^\tau$. Further, let $U = UH \subseteq G$ be an open subset. A smooth function $\varphi : U\tau(U)^{-1} \to \mathbb{C}$ is called τ-*positive definite* if the kernel

$$K(x, y) := \varphi(x\tau(y)^{-1})$$

is positive definite.

Then $\sigma_h(g) := gh$ and $\beta(x)(g) := g.x$ define a smooth right action of $(\mathfrak{g}, H)$ on the manifold U and the kernel K is β-compatible. We therefore obtain for a 1-connected Lie group G^c a corresponding unitary representation U^c on $\mathcal{H}_K \subseteq C^\infty(U, \mathbb{C})$ with

$$(U^c(h)\psi)(g) = \psi(gh) \quad \text{for} \quad g \in U, h \in H$$

and

$$\mathrm{d}U^c(iy)\psi = i\mathcal{L}_y\psi \quad \text{for} \quad \psi \in \mathcal{H}_K^\infty, y \in \mathfrak{q}.$$

So far we worked with scalar-valued kernels, but the corresponding results easily extend to the operator-valued setting as follows:

Example 6.2.6 (Passage to operator-valued kernels) Let (G, τ) be a symmetric Lie group and $g^\sharp = \tau(g)^{-1}$. We consider a smooth right action of G on the manifold X, a complex Hilbert space V, and a hermitian kernel $Q : X \times X \to B(V)$. Further, suppose that $J : G \times X \to GL(V), (g, x) \mapsto J_g(x)$ satisfies the cocycle condition

$$J_{g_1 g_2}(x) = J_{g_1}(x)J_{g_2}(x.g_1) \quad \text{for} \quad g_1, g_2 \in G, x \in X,$$

so that $(g.f)(x) := J_g(x)f(x.g)$ defines a representation of G on V^X. We also assume that the kernel satisfies the corresponding invariance relation

$$J_g(x)Q(x.g, y) = Q(x, y.g^\sharp)J_{g^\sharp}(y)^* \quad \text{for} \quad x, y \in X, g \in G$$

(cf. [Nel64, Proposition II.4.3]). On the set $M := X \times V$, we then obtain a G-right action by

$$(x, v).g := (x.g, J_g(x)^*v).$$

We also obtain a positive definite kernel

$$K : M \times M \to \mathbb{C}, \quad K((x, v), (y, w)) := \langle v, Q(x, y)w \rangle$$

which satisfies the natural covariance condition

$$K((x, v).g, (y, w)) = K((x.g, J_g(x)^*v), (y, w)) = \langle J_g(x)^*v, Q(x.g, y)w \rangle$$
$$= \langle v, J_g(x)Q(x.g, y)w \rangle = \langle v, Q(x, y.g^\sharp)J_{g^\sharp}(y)^*w \rangle$$
$$= K((x, v), (y.g^\sharp, J_{g^\sharp}(y)^*w)) = K((x, v), (y, w).g^\sharp).$$

Let $X_+ \subseteq X$ be an open H-invariant subset on which the kernel Q is positive definite, so that K is positive definite on $M_+ := X_+ \times V$. The corresponding reproducing kernel Hilbert space $\mathcal{H}_K \subseteq \mathbb{C}^{M_+}$ consists of functions that are continuous anti-linear in the second argument, and it is easy to see that the map

$$\Gamma : \mathcal{H}_Q \to \mathcal{H}_K, \quad \Gamma(f)(x, v) := \langle v, f(x) \rangle$$

is unitary (Example A.1.4). In view of

$$\Gamma(g.f)(x, v) = \langle v, J_g(x)f(x.g) \rangle = \langle J_g(x)^*v, f(x.g) \rangle = \Gamma(f)((x, v).g),$$

it intertwines the representation of G on V^X with the action on $\mathbb{C}^M$ by

$$(g.F)(x, v) := F((x, v).g).$$

Assume that the G-action on the Banach manifold M is smooth, i.e., that the map $G \times X \times V \to V, (g, x, v) \mapsto J_g(x)^*v$ is smooth. Then we obtain a smooth right action of $(\mathfrak{g}, H)$ on M_+ compatible with the kernel K, and thus Theorem 6.2.3 yields a unitary representation of G^c on the Hilbert $\mathcal{H}_K \cong \mathcal{H}_Q$.

6.3 Representations on Spaces of Distributions

Now we slightly change our context. To extend the theory from smooth kernels to distribution kernels, we assume that M is a finite dimensional smooth manifold and that $K \in C^{-\infty}(M \times M)$ is a positive definite distribution so that $\mathcal{H}_K \subseteq C^{-\infty}(M)$ holds for the corresponding reproducing kernel Hilbert space (Sect. 2.4). The canonical map

$$\iota : C_c^\infty(M) \to \mathcal{H}_K, \quad \varphi \mapsto K_\varphi$$

is continuous ([MNO15, §7.1]) and therefore the kernel K defines a continuous hermitian form on $C_c^\infty(M)$. Hence Corollary 6.1.2 applies in particular to K.

Definition 6.3.1 The Lie derivative defines on $C_c^\infty(M)$ the structure of a $\mathcal{V}(M)$-module, and we consider on $C^{-\infty}(M)$ the adjoint representation:

$$(\mathcal{L}_X D)(\varphi) := -D(\mathcal{L}_X \varphi) \quad \text{for} \quad X \in \mathcal{V}(M), D \in C^{-\infty}(M), \varphi \in C_c^\infty(M).$$

For a distribution $D \in C^{-\infty}(M \times M)$ and $X \in \mathscr{V}(M)$, we write

$$(\mathscr{L}_X^1 D)(\varphi \otimes \psi) := -D(\mathscr{L}_X \varphi \otimes \psi) \quad \text{and} \quad (\mathscr{L}_X^2 D)(\varphi \otimes \psi) := -D(\varphi \otimes \mathscr{L}_X \psi).$$

We say that X is *D-symmetric, resp., D-skew-symmetric* if $\mathscr{L}_X^1 D = \varepsilon \mathscr{L}_X^2 D$ for $\varepsilon = 1$, resp., -1.

Remark 6.3.2 Let K be a positive definite distribution on M. If X is K-symmetric (resp. K-skew-symmetric), then $\mathscr{L}_X$ defines a symmetric (resp. skew-symmetric) operator on $C_c^\infty(M)$ with respect to $\langle \cdot, \cdot \rangle_K$.

The next observation allows us to use Corollary 6.1.2 and to adapt the methods used in Sect. 6.2.

Remark 6.3.3 Let $X \in \mathscr{V}(M)$ and $\varphi \in C_c^\infty(M)$. We write $M_t \subseteq M$ for the open subset of all points $m \in M$ for which the integral curve of X through m is defined in $t \in \mathbb{R}$. The corresponding time t flow map is denoted $\Phi_t^X : M_t \to M$. If $\operatorname{supp} \varphi \subseteq M_{-t}$, then $\varphi \circ \Phi_t^X$ has compact support $\Phi_{-t}^X(\operatorname{supp} \varphi) \subseteq M_t$ and therefore defines an element of $C_c^\infty(M)$.

Theorem 6.3.4 (Geometric Fröhlich Theorem for distributions) *Let M be a smooth manifold and $K \in C^{-\infty}(M \times M)$ be a positive definite distribution. If $X \in \mathscr{V}(M)$ is K-symmetric, then the Lie derivative $\mathscr{L}_X$ defines an essentially selfadjoint operator $\mathscr{H}_K^0 \to \mathscr{H}_K$ whose closure $\mathscr{L}_X^K$ coincides with $\mathscr{L}_X|_{\mathscr{D}_X}$, where*

$$\mathscr{D}_X := \{D \in \mathscr{H}_K : \mathscr{L}_X D \in \mathscr{H}_K\}.$$

If the local flow Φ^X is defined on $[0, \varepsilon] \times \operatorname{supp}(\varphi)$ for some $\varphi \in C_c^\infty(M)$, then

$$e^{t \mathscr{L}_X^K} K_\varphi = K_{\varphi \circ \Phi_{-t}^X} \quad \text{for} \quad 0 \le t \le \varepsilon. \tag{6.3}$$

Proof For every $\varphi \in C_c^\infty(M)$, there exists an $\varepsilon > 0$ such that the flow Φ^X of X is defined on the compact subset $[0, \varepsilon] \times \operatorname{supp}(\varphi)$ of $\mathbb{R} \times M$. Then the curve

$$\gamma [0, \varepsilon] \to C_c^\infty(M), \quad \gamma(t) := \varphi \circ \Phi_{-t}^X$$

satisfies $\gamma'(t) = -\mathscr{L}_X \varphi$ in the natural topology on $C_c^\infty(M)$. Therefore the assumptions of Corollary 6.1.2 are satisfied with $V = C_c^\infty(M)$ and $L = -\mathscr{L}_X$. We conclude that $\mathscr{L}_X|_{\mathscr{H}_K^0}$ is essentially selfadjoint with closure equal to $\mathscr{L}_X^K$ and that (6.3) holds. $\qquad \square$

Definition 6.3.5 Let $\mathfrak{g} = \mathfrak{h} + \mathfrak{q}$ be a symmetric Lie algebra with involution τ and $\beta : \mathfrak{g} \to \mathscr{V}(M)$ be a homomorphism of Lie algebras. A positive definite distribution $K \in C^{-\infty}(M \times M, \mathbb{C})$ is said to be *β-compatible* if

$$\mathscr{L}^1_{\beta(x)} K = -\mathscr{L}^2_{\beta(\tau(x))} K \quad \text{for} \quad x \in \mathfrak{g}.$$

In the following we assume that K is a positive definite distribution on M compatible with the smooth right action (β, σ) of $(\mathfrak{g}, H)$ (cf. Definition 6.2.2). For $z = x + iy \in \mathfrak{g}_{\mathbb{C}}$, we put

$$\mathscr{L}_{\beta(z)} := \mathscr{L}_{\beta(x)} + i\mathscr{L}_{\beta(y)}$$

and we write $\mathscr{L}_z$ for the restriction of $\mathscr{L}_{\beta(z)}$ to its maximal domain

$$\mathscr{D}_z = \{D \in \mathscr{H}_K : \mathscr{L}_{\beta(z)} D \in \mathscr{H}_K\}.$$

As in Sect. 6.2, we consider the subspace

$$\mathscr{D} := \{D \in \mathscr{H}_K : (\forall n \in \mathbb{N})(\forall x_1, \ldots, x_n \in \mathfrak{g}) \, \mathscr{L}_{\beta(x_1)} \cdots \mathscr{L}_{\beta(x_n)} D \in \mathscr{H}_K\}$$

which carries the Lie algebra representation $\alpha \, \mathfrak{g}_{\mathbb{C}} \to \text{End}(\mathscr{D})$ for which the operator $\alpha(x)$ is skew-hermitian for $x \in \mathfrak{g}^c = \mathfrak{h} + i\mathfrak{q}$. From (6.1) and Remark 6.3.2 we deduce that

$$\mathscr{L}_x K_\varphi = K_{\mathscr{L}_{\tau(x)}\varphi} \quad \text{for} \quad \varphi \in C_c^\infty(M), \tag{6.4}$$

hence $\mathscr{H}_K^0 \subseteq \mathscr{D}$. In particular, $\mathscr{D}$ is dense in $\mathscr{H}_K$.

Theorem 6.3.6 *Let $K \in C^{-\infty}(M \times M)$ be a positive definite distribution compatible with the smooth right action (β, σ) of the pair $(\mathfrak{g}, H)$ on M, where $\mathfrak{g} = \mathfrak{h} \oplus \mathfrak{q}$ is a symmetric Lie algebra and H is a connected Lie group with Lie algebra $\mathfrak{h}$. Let G^c be a simply connected Lie group with Lie algebra $\mathfrak{g}^c = \mathfrak{h} + i\mathfrak{q}$. Then there exists a unique smooth unitary representation $(U^c, \mathscr{H}_K)$ of G^c such that*

$$\overline{dU^c(x)} = \mathscr{L}_x \quad \text{and} \quad \overline{dU^c(iy)} = i\mathscr{L}_y \quad \text{for} \quad x \in \mathfrak{h}, y \in \mathfrak{q}.$$

6.4 Reflection Positive Distributions and Representations

In this subsection we connect the previously described integrability results to reflection positive representations. Let $D \in C^{-\infty}(M \times M, \mathbb{C})$ be a positive definite distribution which is reflection positive with respect to the involution $\theta : M \to M$ on the open subset $M_+ \subseteq M$ (cf. Definition 2.4.5). Our main result is Theorem 6.4.1 which asserts that, under the natural compatibility requirements for an action of a symmetric Lie group (G, H, τ) on (M, θ), the representation of the pair $(\mathfrak{g}^c, H)$ on the Hilbert space $\mathscr{H}_{D^\theta}$ corresponding to the positive definite distribution kernel D^θ on M_+ integrates to a unitary representation of the simply connected group G^c with Lie algebra $\mathfrak{g}^c$.

Let (G, H, τ) be a symmetric Lie group acting on M such that $\theta(g.m) = \tau(g).\theta(m)$ and $H.M_+ = M_+$. We assume that D is invariant under G and τ. Then we have a natural unitary representation $(U_{\mathscr{E}}, \mathscr{E})$ of G on the Hilbert subspace $\mathscr{E} := \mathscr{H}_K \subseteq C^{-\infty}(M)$. As M_+, and therefore $\mathscr{E}_+$, is H-invariant, this representation is infinitesimally reflection positive in the sense of Definition 3.3.6.

From the invariance condition

$$\mathscr{L}^1_{\beta(x)}D = -\mathscr{L}^2_{\beta(x)}D \quad \text{for } x \in \mathfrak{g} \tag{6.5}$$

we derive

$$\mathscr{L}^1_{\beta(x)}D^\theta = -\mathscr{L}^2_{\beta(\tau(x))}D^\theta \quad \text{for } x \in \mathfrak{g}. \tag{6.6}$$

This implies that the assumptions of Theorem 6.3.6 are satisfied, so that we obtain:

Theorem 6.4.1 *Let M be a smooth finite dimensional manifold and $D \in C^{-\infty}(M \times M)$ be a positive definite distribution which is reflection positive with respect to (M, M_+, θ). Let (G, H, τ) be a symmetric Lie group acting on M such that $\theta(g.m) = \tau(g).\theta(m)$ and $H.M_+ = M_+$. We assume that D is invariant under G and τ. Let G^c be a simply connected Lie group with Lie algebra $\mathfrak{g}^c = \mathfrak{h} + i\mathfrak{q}$ and define $(\mathscr{L}_x)_{x \in \mathfrak{g}}$ on its maximal domain in the Hilbert subspace $\mathscr{H}_{D^\theta} \subseteq C^{-\infty}(M_+)$. Then there exists a unique unitary representation $(U^c, \mathscr{H}_{D^\theta})$ of G^c such that*

$$\overline{\mathrm{d}U^c(x)} = \mathscr{L}_x \quad \text{and} \quad \overline{\mathrm{d}U^c(iy)} = i\mathscr{L}_y \quad \text{for } x \in \mathfrak{h}, y \in \mathfrak{q}.$$

Example 6.4.2 The preceding theorem applies in particular to the situation where $M = G_\tau$, $\tau(g) = g\tau$ and $M_+ = G_+$ is an open subset of G with $G_+ H = G_+$ (Remark 3.4.2). Here we start with a reflection positive distribution $D \in C^{-\infty}(G_\tau)$ (Definition 7.2.1). It defines a G_τ-invariant distribution kernel $\widetilde{D}$ on $G_\tau \times G_\tau$ which is reflection positive with respect to G_+. We thus obtain a positive definite distribution $\widetilde{D}^\tau$ on $G_+ \times G_+$.

Example 6.4.3 Reflection positive representations of the euclidean motion group $E(d)$ (cf. Example 3.2.2) lead to unitary representations of the simply connected covering $G^c = \mathbb{R}^d \rtimes \mathrm{Spin}_{1,d-1}(\mathbb{R})$ of the identity component $P(d)_0$ of the Poincaré group. More concretely, we consider $M = \mathbb{R}^d$, $M_+ = \mathbb{R}^d_+$, $\tau(x_0, \mathbf{x}) = (-x_0, \mathbf{x})$ and $G = E(d) = \mathbb{R}^d \rtimes O_d(\mathbb{R})$. Then the G-invariance of a distribution $\widetilde{D}$ on $\mathbb{R}^d \times \mathbb{R}^d$ means that it is determined by an $O_d(\mathbb{R})$-invariant distribution $D \in C^{-\infty}(\mathbb{R}^d)$ by

$$\widetilde{D}(\varphi \otimes \psi) := D(\varphi^\vee * \psi), \qquad \varphi^\vee(x) := \varphi(-x).$$

For any reflection positive rotation invariant distribution $D \in C^{-\infty}(\mathbb{R}^d)$, we thus obtain a reflection positive representation $(U_{\mathscr{E}}, \mathscr{E})$ of G and a representation of the group G^c on $\widehat{\mathscr{E}} \cong \mathscr{H}_{D^\theta}$.

For $d \geq 3$, the natural inclusion $\mathrm{SO}_{d-1}(\mathbb{R}) \to O_{1,d-1}(\mathbb{R})$, $g \mapsto \mathrm{id}_\mathbb{R} \times g$ induces a surjective homomorphism $\pi_1(\mathrm{SO}_{d-1}(\mathbb{R})) \to \pi_1(O_{1,d-1}(\mathbb{R}))$, and since U^c is compatible with the unitary representation $\widehat{U}^H$ of H on $\widehat{\mathcal{E}}$, the representation U^c factors through a representation of the connected Poincaré group $P(d)_0 = \mathbb{R}^d \rtimes \mathrm{SO}_{1,d-1}(\mathbb{R})$.

The concrete case of generalized free fields discussed in Chap. 8 is of basic interest.

Notes

Main reference for this section is [MNO15], where the results on smooth kernels are developed in the more general context of Banach–Lie groups acting on locally convex manifolds. Here we choose the simpler context of Banach manifolds because in this context every smooth vector field has a local flow.

Fröhlich's results from [Fro80] have later been refined in several ways, in particular by Klein and Landau in [KL81, KL82]. Fröhlich, Osterwalder and Seiler introduced in [FOS83] the concept of a virtual representation, which was developed in greater generality by Jorgensen in [Jo86, Jo87].

In the context of involutive representations of a subsemigroup $S \subseteq G$ with polar decomposition $S = H \exp C$, where $C \neq \emptyset$ is an $\mathrm{Ad}(H)$-invariant open convex cone in q, the Lüscher–Mack Theorem [LM75, HN93, MN12] provides a corresponding unitary representation of the dual group G^c.

Chapter 7
Reflection Positive Distribution Vectors

In this chapter we first introduce the concept of a distribution vector of a unitary representation (Sect. 7.1). It turns out that certain distribution vectors semi-invariant under a subgroup H correspond naturally to realizations of the representation in a Hilbert space of distributions on the homogeneous space G/H. In this context reflection positive representations can be constructed from reflection positive distributions on G/H (Sect. 7.2). Such distributions can often be found and even classified in terms of the geometry of the homogeneous space.

To illustrate this technique, we apply it in Sect. 7.3 to spherical representations of the Lorentz group $G = O_{1,n}(\mathbb{R})$. These representations consist of two series, the principal series and the complementary series. Both have natural realizations in spaces of distributions on the n-sphere $\mathbb{S}^n \cong G/P$ on which the Lorentz group G acts by conformal maps; the principal series can even be realized in $L^2(\mathbb{S}^n)$. That some of the representation of the complementary series exhibit reflection positivity with respect to the subsemigroup of conformal compressions of a half-sphere is shown in Sect. 7.4.1. In Sect. 7.4.2 we build a bridge from the natural reflection positivity on the sphere $\mathbb{S}^n$ as a Riemannian manifold obtained from resolvents $(m^2 - \Delta)^{-1}$ of the Laplacian (cf. Sect. 2.5) and unitary representations. Here the Lorentz group occurs as the dual group G^c of the isometry group $G = O_{n+1}(\mathbb{R})$ of $\mathbb{S}^n$ and we identify the unitary representations of G^c on the corresponding Hilbert spaces $\widehat{\mathscr{E}}$ as spherical representations of G^c realized in a space of holomorphic functions in the crown domain of hyperbolic space.

7.1 Distribution Vectors

In this subsection we introduce the notion of distribution vectors and in the following section we connect it with reflection positivity. We start with the basic structures related to distributions on Lie groups and homogeneous spaces.

© The Author(s) 2018
K.-H. Neeb and G. Ólafsson, *Reflection Positivity*, SpringerBriefs in Mathematical Physics, https://doi.org/10.1007/978-3-319-94755-6_7

7.1.1 Distributions on Lie Groups and Homogeneous Spaces

Definition 7.1.1 Let G be a Lie group. We fix a left invariant Haar measure μ_G on G. This measure defines on $L^1(G) := L^1(G, \mu_G)$ the structure of a Banach-$*$-algebra by the *convolution product*

$$(\varphi * \psi)(u) = \int_G \varphi(g)\psi(g^{-1}u)\,d\mu_G(g), \quad \text{and} \quad \varphi^*(g) = \overline{\varphi(g^{-1})}\Delta_G(g)^{-1} \quad (7.1)$$

is the involution, where $\Delta_G : G \to \mathbb{R}_+$ is the *modular function* determined by

$$\int_G \varphi(y)\,d\mu_G(y) = \int_G \varphi(y^{-1})\Delta_G(y)^{-1}\,d\mu_G(y) \quad \text{and}$$

$$\Delta_G(x)\int_G \varphi(yx)\,d\mu_G(y) = \int_G \varphi(y)\,d\mu_G(y) \quad \text{for} \quad \varphi \in C_c(G).$$

The formulas above show that we have two isometric actions of G on $L^1(G)$, given by

$$(\lambda_g f)(x) = f(g^{-1}x) \quad \text{and} \quad (\rho_g f)(x) = f(xg)\Delta_G(g). \qquad (7.2)$$

Note that $(\lambda_g f)^* = \rho_g f^*$.

Let $H \subseteq G$ be a closed subgroup and $X := G/H = \{gH : g \in G\}$ be the space of H-left cosets, endowed with its canonical manifold structure. Let μ_H denote a left Haar measure on H. Then the map

$$\alpha \colon C_c^\infty(G) \to C_c^\infty(X), \ \varphi \mapsto \varphi^\flat, \quad \varphi^\flat(gH) := \int_H \varphi(gh)\,d\mu_H(h)$$

is a topological quotient map, i.e., surjective, continuous and open (cf. [Wa72, p. 475] and [vD09, p. 136]). Its adjoint thus provides an injection

$$\alpha^* \colon C^{-\infty}(X) \hookrightarrow C^{-\infty}(G), \ D \mapsto D^\sharp, \quad D^\sharp(\varphi) := D(\varphi^\flat)$$

of the space of distributions on $X = G/H$ into the space of distributions on G.

On $C_c^\infty(X)$ the group G acts naturally by left translations $(\lambda_g\varphi)(x) := \varphi(g^{-1}x)$ and, accordingly, by $(\lambda_g D)(\varphi) := D(\lambda_g^{-1}\varphi)$ on distributions. We also recall the two G-actions (7.2) on $C_c^\infty(G) \subseteq L^1(G)$ by left and right translations and note that they induce actions on the dual space $C^{-\infty}(G)$. As

$$\alpha \circ \lambda_g = \lambda_g \circ \alpha \quad \text{and} \quad \alpha \circ \rho_h = \Delta_G(h)\Delta_H(h)^{-1}\alpha \quad \text{for} \quad h \in H, g \in G, \quad (7.3)$$

the map α and its adjoint intertwine the left translation actions of G on $C^{-\infty}(X)$ and $C^{-\infty}(G)$. It also follows that $\rho_h\alpha^*(D)(\varphi) = D(\alpha(\rho_h^{-1}\varphi)) = \frac{\Delta_H(h)}{\Delta_G(h)}\alpha^*(D)(\varphi)$, and we even have:

Lemma 7.1.2 *The range of* α^* *is the space* $C^{-\infty}(G)_H$ *of all distributions* $D \in$ $C^{-\infty}(G)$ *satisfying*

$$\rho_h D = \Delta_{G/H}(h)^{-1} D \quad \text{with} \quad \Delta_{G/H}(h) := \frac{\Delta_G(h)}{\Delta_H(h)} \quad \text{for} \quad h \in H. \tag{7.4}$$

Proof Let $D \in C^{-\infty}(G)_H$. As α is a quotient map, we have to show that $\ker \alpha \subseteq$ $\ker D$ because $\mathrm{im}(\alpha)^* = (\ker \alpha)^\perp$. First we note that, for $\psi \in C_c^\infty(G)$, the distribution defined by $(\psi * D)(\varphi) := D(\psi^* * \varphi)$ has a smooth density Ψ with respect to μ_G satisfying $\rho_h \Psi = \Delta_{G/H}(h)^{-1} \Psi$ for $h \in H$.

Let $\rho : G \to (0, \infty)$ denote a smooth function on G with

$$\rho(e) = 1 \quad \text{and} \quad \rho(gh) = \rho(g)\Delta_{G/H}(h)^{-1} \quad \text{for} \quad g \in G, h \in H,$$

and write $\mu_{G/H}$ for the corresponding quasi-invariant measure on G/H defined by

$$\int_{G/H} \varphi^\flat(gH) \, d\mu_{G/H}(gH) = \int_G \varphi(g)\rho(g) \, d\mu_G(g)$$

([Wa72, p. 475]). That $\mu_{G/H}$ is well defined requires to verify that $\varphi^\flat = 0$ implies that the right hand side vanishes. Now

$$(\psi * D)(\varphi) = D(\psi^* * \varphi) = \int_G \overline{\varphi(g)} \Psi(g) \, d\mu_G(g) = \int_{G/H} (\overline{\varphi} \Psi \rho^{-1})^\flat (gH) \, d\mu_{G/H}(gH)$$

$$= \int_{G/H} \overline{\varphi}^\flat(gH)(\Psi \rho^{-1})(g) \, d\mu_{G/H}(gH) = 0.$$

Replacing ψ by a δ-sequence in $C_c^\infty(G)$, we obtain for $n \to \infty$ that $D(\varphi) = 0$. □

The distribution $D_{\mu_G}(\varphi) := \int_G \overline{\varphi(g)} \, d\mu_G(g)$ is left and right invariant, hence contained in $C^{-\infty}(G)_H$ if and only if $\Delta_{G/H} = 1$. If this is the case, then $D_{\mu_G} = D_{\mu_X}^\sharp$ for a G-invariant measure μ_X on X. One can even show that, conversely, the existence of such a measure implies the vanishing of $\Delta_{G/H}$ (cf. [Wa72] or [HN12, Sect. 10.4]).

7.1.2 Smooth Vectors and Distribution Vectors

Now let $(U, \mathcal{H})$ be a unitary representation of the Lie group G, i.e., a homomorphism $U : G \to \mathrm{U}(\mathcal{H})$, $g \mapsto U_g$, such that for each $\eta \in \mathcal{H}$ the orbit map $U^\eta(g) = U_g \eta$ is continuous. We say that $\eta \in \mathcal{H}$ is *smooth* if $U^\eta : G \to \mathcal{H}$ is smooth. The space of smooth vectors is denoted by $\mathcal{H}^\infty$. This space carries a representation dU of the Lie algebra $\mathfrak{g}$ on $\mathcal{H}^\infty$ given by

$$\mathrm{d}U(x)\eta = \lim_{t\to 0} \frac{U_{\exp tx}\eta - \eta}{t}.$$

For a basis $x_1, \ldots, x_k$ of $\mathfrak{g}$ and $\mathbf{m} = (m_1, \ldots, m_k) \in \mathbb{N}_0^k$ the family of semi-norms

$$q_{\mathbf{m}}(\eta) = \|\mathrm{d}U(x_1)^{m_1} \cdots \mathrm{d}U(x_k)^{m_k}\eta\|$$

defines a Fréchet space topology on $\mathscr{H}^\infty$ such that the inclusion $\mathscr{H}^\infty \hookrightarrow \mathscr{H}$ is continuous. The space $\mathscr{H}^\infty$ is G-invariant and $\mathrm{d}U(\mathrm{Ad}(g)x) = U_g \mathrm{d}U(x)U_g^{-1}$.

For $\varphi \in L^1(G)$ the operator-valued integral $U_\varphi := \int_G \varphi(g)U_g \, d\mu_G(g)$ exists and is uniquely determined by

$$\langle \eta, U_\varphi \zeta \rangle = \int_G \varphi(g)\langle \eta, U_g \zeta \rangle \, d\mu_G(g) \quad \text{for} \quad \eta, \zeta \in \mathscr{H}. \tag{7.5}$$

Then $\|U_\varphi\| \le \|\varphi\|_1$ and the so-obtained continuous linear map $L^1(G) \to B(\mathscr{H})$ is a representation of the Banach-$*$-algebra $L^1(G)$, i.e., $U_{\varphi*\psi} = U_\varphi U_\psi$ and $U_{\varphi^*} = U_\varphi^*$. Note that $U_g U_\varphi = U_{\lambda_g \varphi}$ and $U_\varphi U_g = U_{\rho_g \varphi}$.

The space $\mathscr{H}^\infty$ of smooth vectors is G-invariant and we denote the corresponding representation by U^∞. If $\varphi \in C_c^\infty(G)$ and $\xi \in \mathscr{H}$, then $U_\varphi \xi \in \mathscr{H}^\infty$ and

$$\mathrm{d}U(x)U_\varphi \xi := U_{\mathrm{d}\lambda_x \varphi}\xi, \quad \text{where} \quad \mathrm{d}\lambda_x \varphi = \frac{d}{dt}\Big|_{t=0} \lambda_{\exp tx}\varphi.$$

This follows directly by differentiation under the integral sign. If $(\varphi_n)_{n\in\mathbb{N}}$ is a δ-sequence, then $U_{\varphi_n}\xi \to \xi$, so that $\mathscr{H}^\infty$ is dense in $\mathscr{H}$.

The space of continuous anti-linear functionals on $\mathscr{H}^\infty$ is denoted by $\mathscr{H}^{-\infty}$. Its elements are called *distribution vectors*. The group G and its Lie algebra $\mathfrak{g}$ act on $\mathscr{H}^{-\infty}$ by

$$(U_g^{-\infty}\eta)(\xi) := \eta(U_{g^{-1}}^\infty \xi), \quad \text{resp.,} \quad (\mathrm{d}U^{-\infty}(x)\eta)(\xi) := -\eta(\mathrm{d}U(x)\xi), \quad g \in G, x \in \mathfrak{g}.$$

We then have $U_\varphi^{-\infty}\eta := \eta \circ U_{\varphi^*}^\infty$ for $\varphi \in C_c^\infty(G)$. We obtain natural G-equivariant linear embeddings

$$\mathscr{H}^\infty \xrightarrow{\ \iota_\infty(\xi)=\xi\ } \mathscr{H} \xrightarrow{\ \iota_{-\infty}(\xi)=\langle\cdot,\xi\rangle\ } \mathscr{H}^{-\infty}$$

and note that $U_\varphi^{-\infty}\mathscr{H}^{-\infty} \subseteq \mathscr{H}^\infty$ for $\varphi \in C_c^\infty(G)$.

Example 7.1.3 Let $H \subseteq G$ be a closed subgroup and $X = G/H$. Then there exists a quasi-invariant measure μ_X on X with a smooth density with respect to Lebesgue measure in any chart; for details see [Fo95, Sec. 2.6], [Fa00, p. 146ff] and [HN12, Sect. 10.4]. Thus there exists a smooth strictly positive function $j : G \times X \to \mathbb{R}_+$ such that for all $\varphi \in C_c^\infty(X)$ and $g \in G$ we have $\int_X \varphi(g.x)\,d\mu_X(x) = \int_X \varphi(x)j(g,x)\,d\mu_X(x)$ or, equivalently,

$$\int_X \varphi(g^{-1}.x)\,j(g,x)\,d\mu_X(x) = \int_X \varphi(x)\,d\mu_X(x) \quad \text{for} \quad \varphi \in C_c^\infty(X). \quad (7.6)$$

We will also write $j_g(x) = j(g,x)$. As a consequence of (7.6), we obtain a family of unitary representation, the *quasi-regular representation* of G on $L^2(X)$ by

$$U_g^\lambda \varphi(x) := j(g,x)^{\lambda+1/2}\varphi(g^{-1}.x) \quad \text{for} \quad \lambda \in i\mathbb{R}, g \in G \text{ and } \varphi \in L^2(X). \quad (7.7)$$

If $(U, \mathscr{H})$ is a unitary representation of G and $\eta \in \mathscr{H}^{-\infty}$, then the adjoint of the linear map $C_c^\infty(G) \to \mathscr{H}^\infty$, $\varphi \mapsto U_\varphi^{-\infty}\eta$, defines a linear map

$$j_\eta: \mathscr{H}^{-\infty} \to C^{-\infty}(G), \quad j_\eta(\alpha)(\varphi) := \alpha(U_\varphi^{-\infty}\eta).$$

For $g \in G$, we have $j_\eta \circ U_g^{-\infty} = \lambda_g \circ j_\eta$, i.e., j_η intertwines the action of G on $\mathscr{H}^{-\infty}$ with the left translation action on $C^{-\infty}(G)$.

We now introduce the concepts used in the proposition below.

Definition 7.1.4 We call the distribution vector $\eta \in \mathscr{H}^{-\infty}$ cyclic if $U_{C_c^\infty(G)}^{-\infty}\eta$ is dense in $\mathscr{H}$.

Definition 7.1.5 We call a distribution $D \in C^{-\infty}(G)$ *positive definite* if the sesquilinear kernel

$$K_D(\varphi, \psi) := D(\psi^* * \varphi) \quad \text{on} \quad C_c^\infty(G) \quad (7.8)$$

is positive semidefinite. This is equivalent to the positive definiteness of the distribution $\tilde{D}$ on $G \times G$ determined by

$$\tilde{D}(\overline{\psi} \otimes \varphi) = D(\psi^* * \varphi) \quad \text{for} \quad \varphi, \psi \in C_c^\infty(G).$$

We also note that (7.8) is equivalent to D defining a positive functional on the $*$-algebra $C_c^\infty(G)$, endowed with the convolution product. The corresponding reproducing kernel Hilbert space $\mathscr{H}_D := \mathscr{H}_{K_D}$ is a linear subspace of $C^{-\infty}(G)$ in which the distributions defined by $\psi * D = \lambda_\psi(D)$, i.e., $(\psi * D)(\varphi) := D(\psi^* * \varphi)$, form a dense subspace with

$$\langle \varphi * D, \psi * D \rangle = D(\psi^* * \varphi) \quad \text{for} \quad \varphi, \psi \in C_c^\infty(G). \quad (7.9)$$

In particular, $D \in \mathscr{H}_D^{-\infty}$, $j_D(D) = D$ and $j_D|_{\mathscr{H}_D}: \mathscr{H}_D \hookrightarrow C^{-\infty}(G)$ is the inclusion map.

Proposition 7.1.6 (Realization in spaces of distributions) *Let $(U, \mathscr{H})$ be a unitary representation of G and $\eta \in \mathscr{H}^{-\infty}$. Then the following assertions hold:*

(a) *The map $j_\eta: \mathscr{H}^{-\infty} \to C^{-\infty}(G)$ is injective if and only if η is cyclic.*
(b) *The distribution $D_\eta := j_\eta(\eta)$ is positive definite.*

(c) *If η is cyclic, then $j_\eta : \mathscr{H} \to \mathscr{H}_\eta := j_\eta(\mathscr{H}) \subset C^{-\infty}(G)$ is a G-invariant unitary operator onto the reproducing kernel Hilbert space of distribution on G for which the inner product and the reproducing kernel are determined by*

$$\langle j_\eta(U_\varphi^{-\infty}\eta), j_\eta(U_\psi^{-\infty}\eta)\rangle_{\mathscr{H}_\eta} = \langle U_\varphi^{-\infty}\eta, U_\psi^{-\infty}\eta\rangle_{\mathscr{H}} = D_\eta(\psi^* * \varphi).$$

(d) *If $H \subset G$ is a closed subgroup, then $j_\eta(\mathscr{H}^{-\infty}) \subseteq C^{-\infty}(G)_H$ if and only if $U_h^{-\infty}\eta = \Delta_{G/H}(h)\eta$ holds for all $h \in H$.*

Proof For (a), we first observe that the injectivity of j_η on $\mathscr{H}$ is trivially equivalent to η being cyclic. To see that this even implies that j_η is injective on $\mathscr{H}^{-\infty}$, assume that $j_\eta(\alpha) = 0$. Then equivariance implies $j_\eta(U_\varphi^{-\infty}\alpha) = 0$ for every $\varphi \in C_c^\infty(G)$. For any δ-sequence $(\delta_n)_{n\in\mathbb{N}}$ in $C_c^\infty(G)$, we have $U_{\delta_n}^{-\infty}\alpha \to \alpha$ in the weak-$*$-topology on $\mathscr{H}^{-\infty}$, and since j_η is obviously weak-$*$-continuous, it follows that $j_\eta(\alpha) = 0$.

For (b) we derive from $D_\eta(\varphi) = \eta(U_\varphi^{-\infty}\eta)$ the relation

$$D_\eta(\psi^* * \varphi) = \eta(U_{\psi^* * \varphi}^{-\infty}\eta) = \eta(U_{\psi^*}^\infty U_\varphi^{-\infty}\eta) = (U_\psi^{-\infty}\eta)(U_\varphi^{-\infty}\eta) = \langle U_\varphi^{-\infty}\eta, U_\psi^{-\infty}\eta\rangle.$$

By Remark A.1.2, this implies that D_η is positive definite.

(c) follows from the fact that, for $\varphi, \psi \in C_c^\infty(G)$,

$$j_\eta(U_\psi^{-\infty}\eta)(\varphi) = \langle U_\varphi^{-\infty}\eta, U_\psi^{-\infty}\eta\rangle = D_\eta(\psi^* * \varphi) = (\psi * D_\eta)(\varphi).$$

To obtain (d), we first observe the relation $j_\eta(\alpha)(\varphi) = \overline{j_\alpha(\eta)(\varphi^*)}$ for $\alpha, \eta \in \mathscr{H}^{-\infty}$, which easily follows from the existence of a factorization $\varphi = \varphi_1 * \varphi_2$ with $\varphi_j \in C_c^\infty(G)$ (Dixmier–Malliavin Theorem [DM78, Theorem 3.1]). For $h \in H$, this leads to

$$(\rho_h D_\eta)(\varphi) = D_\eta(\rho_h^{-1}\varphi) = \eta(U_\varphi^{-\infty}U_{h^{-1}}^{-\infty}\eta) = j_{U_{h^{-1}}^{-\infty}\eta}(\eta)(\varphi) = \overline{j_\eta(U_{h^{-1}}^{-\infty}\eta)(\varphi^*)},$$

so that the assertion follows from (7.4). $\square$

From the preceding proposition we derive:

Theorem 7.1.7 *A unitary representation $(U, \mathscr{H})$ can be realized on a Hilbert subspace of $C^{-\infty}(G/H)$ if and only if there exists a cyclic distribution vector $\eta \in \mathscr{H}^{-\infty}$ satisfying $U_h^{-\infty}\eta = \Delta_{G/H}(h)\eta$ for $h \in H$.*

Example 7.1.8 (a) Let G be a Lie group and H a closed subgroup such that $X = G/H$ carries a G-invariant measure μ_X. Then G acts unitarily on $L^2(X) = L^2(X, \mu_X)$ by $\lambda_g\varphi(x) = \varphi(g^{-1}.x)$. The space $L^2(X)^\infty$ of smooth vectors is the space of smooth functions $\varphi \in C^\infty(X)$ such that $\lambda_u f \in L^2(X)$ for all $u \in U(\mathfrak{g})$ ([Po72, Theorem 5.1]). If X is compact, then $L^2(X)^\infty = C^\infty(X)$ and $L^2(X)^{-\infty} = C^{-\infty}(X)$ is the space of distributions on X.

(b) Let $G = \mathrm{Heis}(\mathbb{R}^{2n}) = \mathbb{T} \times \mathbb{R}^n \times \mathbb{R}^n$ be the Heisenberg group acting on $L^2(\mathbb{R}^n)$ via the Schrödinger representation

$$(\pi(z, x, y)f)(u) = ze^{i\langle x,u\rangle} f(u - y)$$

(cf. the proof of Theorem 4.4.1). Then the space $L^2(\mathbb{R}^n)^\infty$ of smooth vectors is $\mathscr{S}(\mathbb{R}^n)$, the Schwartz space of rapidly decreasing functions.

For later reference, we record the following lemma ([NÓ15a, Lemma D.7]) which identifies the distribution vectors in representations of $G = \mathbb{R}^d$ by multiplication operators.

Lemma 7.1.9 *Let $(X, \mathfrak{S}, \mu)$ be a measure space. We write $M(X, \mathbb{C})$ for the vector space of measurable functions $X \to \mathbb{C}$. For $(H_j)_{j=1,\ldots,d}$ in $M(X, \mathbb{R})$ and $R :=$ $\sqrt{\sum_{j=1}^{d} H_j^2}$, we consider the continuous unitary representation of $\mathbb{R}^d$ on $L^2(X, \mu)$, given by*

$$U_{\mathbf{t}}(f) := e^{i \sum_{j=1}^{d} t_j H_j} f \quad \text{for} \quad \mathbf{t} = (t_1, \ldots, t_d).$$

Then

$$\mathscr{H}^{-\infty} \cong \left\{ h \in M(X, \mathbb{C}) \colon (\exists n \in \mathbb{N}) \; \|(1 + R^2)^{-n} f\|_2 < \infty \right\},$$

where the pairing $\mathscr{H}^\infty \times \mathscr{H}^{-\infty} \to \mathbb{C}$ is given by $(f, h) \mapsto \int_X \overline{f} h \, d\mu$. Moreover, the following assertions are equivalent:

(i) *The constant function 1 is a distribution vector.*
(ii) *For the measurable map $\eta := (H_1, \ldots, H_d) \colon X \to \mathbb{R}^d$, the measure $\eta_* \mu$ on $\mathbb{R}^d$ is tempered.*
(iii) *$\widehat{\varphi} \circ \eta \in L^2(X, \mu)$ for every $\varphi \in C_c^\infty(\mathbb{R}^d)$.*

If these conditions are satisfied, then the corresponding distribution on $\mathbb{R}^d$ is given by the Fourier transform of $\eta_* \mu$.

7.2 Reflection Positive Distribution Vectors

In this chapter $(U, \mathscr{E})$ will always denote a unitary representation of a Lie group G and $H \subseteq G$ will be a closed subgroup.

Definition 7.2.1 Let (G, τ) be a symmetric Lie group and $G_\tau = G \rtimes \{\mathrm{id}_G, \tau\}$.

(a) A positive definite τ-invariant distribution $D \in C^{-\infty}(G)$ is called *reflection positive with respect to* (G, G_+, τ) if

$$D(\varphi^\sharp * \varphi) \geq 0 \quad \text{for} \quad \varphi \in C_c^\infty(G_+), \varphi^\sharp(g) := \varphi^*(\tau(g)). \tag{7.10}$$

This is equivalent to the corresponding distribution $\tilde{D}(\overline{\psi} \otimes \varphi) = D(\psi^* * \varphi)$ on $G \times G$ being reflection positive with respect to (G, G_+, τ) (cf. Definition 2.4.5).

(b) Let $(U, \mathscr{E})$ be a unitary representation of G_τ and $\theta := U_\tau$. Then a τ-invariant distribution vector η is said to be *reflection positive* with respect to (G, G_+, τ) if the subspace $\mathscr{E}_+ := \llbracket U^{-\infty}_{C^\infty_c(G_+)} \eta \rrbracket$ is θ-positive (cf. Definition 3.3.1).

(c) If $\eta \in \mathscr{E}^{-\infty}$ is cyclic and reflection positive, then we say that $(U, \mathscr{E}, \eta)$ is a *distribution cyclic reflection positive representation* of G_τ.

For the special case, where $G_+ = S \subseteq G$ holds for an open #-invariant subsemigroup $S \subseteq G$, a positive definite distribution $D \in C^{-\infty}(G)$ is *reflection positive* if $\tau D = D$ and $D|_S$ is positive definite as a distribution on the involutive semigroup $(S, \#)$, i.e., $D(\varphi^\# * \varphi) \geq 0$ for $\varphi \in C^\infty_c(S)$ and $\varphi^\sharp(s) := \varphi^*(\tau(s))$. For a unitary representation of G, a τ-invariant distribution vector $\eta \in \mathscr{E}^{-\infty}$ is reflection positive with respect to S if the subspace $\mathscr{E}_+ := \llbracket U^{-\infty}_{C^\infty_c(S)} \eta \rrbracket$ is θ-positive (cf. Definition 3.4.6).

We now obtain easily:

Theorem 7.2.2 *For (G, G_+, τ) as above, the following assertions holds:*

(a) *If $(U, \mathscr{E}, \eta)$ is a distribution cyclic reflection positive representation of G_τ with respect to (G, G_+, τ), then D_η is reflection positive with respect to (G, G_+, τ).*

(b) *If $D \in C^{-\infty}(G)$ is reflection positive with respect to (G, G_+, τ), then $(U_D, \mathscr{H}_D, D)$ is a distribution cyclic reflection positive representation. If $G_+ = S$ is a $\sharp$-invariant open subsemigroup, then we have an S-equivariant unitary map*

$$\Gamma : \widehat{\mathscr{E}} \to \mathscr{H}_{D|_S} \subset C^{-\infty}(S), \quad \Gamma(\widehat{\varphi * D}) = \varphi|_S.$$

Proof For $\varphi \in C^\infty_c(G_+)$, we have

$$D_\eta(\varphi^\sharp * \varphi) = \eta(U^{-\infty}_{\varphi^\sharp * \varphi} \eta) = \langle U^{-\infty}_{\tau_* \varphi} \eta, U^{-\infty}_\varphi \eta \rangle = \langle \varphi, \varphi \rangle_\theta \geq 0.$$

The other parts of (a), as well as (b), now follow from Lemma 2.4.6. $\square$

7.3 Spherical Representation of the Lorentz Group

In this, and the following section, we discuss reflection positivity related to the conformal geometry of $\mathbb{R}^n$, resp., of its conformal completion $\mathbb{S}^n$. We first discuss the complementary series of the conformal group $G := O_{1,n+1}(\mathbb{R})^\uparrow$ and then we turn to the reflection positivity arising in Riemannian geometry from resolvents of the Laplacian on $\mathbb{S}^n$ as described in Sect. 2.5.

7.3.1 The Principal Series

We write elements of $\mathbb{R}^{n+2}$ as $(x_{-1}, x_0, \mathbf{x})$. Correspondingly, elements of $\mathbb{R}^{n+1}$ are written as $(x_0, \mathbf{x})$, and $e_{-1}, e_0, e_1, \ldots, e_n$ denotes the standard basis of $\mathbb{R}^{n+2}$. We then

identify $e_0, \ldots, e_n$ with the standard basis for $\mathbb{R}^{n+1}$. Elements of G are written as
$g = \begin{pmatrix} a & b^\top \\ c & d \end{pmatrix}$, where $a \in \mathbb{R}$, $b, c \in \mathbb{R}^{n+1}$ and $d \in M_{n+1}(\mathbb{R})$. Recall also the notation
$[x, y] = x_{-1} y_{-1} - \langle x, y \rangle$ and consider the set

$$\mathbb{L}_+^{n+1} := \{x \in \mathbb{R}^{n+2} : [x, x] = 0, x_{-1} > 0\}$$

of positive lightlike vectors. The embedding $\xi : \mathbb{S}^n \to \mathbb{L}_+^{n+1}$, $x \mapsto (1, x)$, yields a
diffeomorphism $\mathbb{S}^n \to \mathbb{L}_+^{n+1} / \mathbb{R}_+^\times$. As the standard linear action of G on $\mathbb{R}^{n+2}$ leaves
$\mathbb{L}_+^{n+1}$-invariant, we thus obtain a smooth action on the quotient space $\mathbb{L}_+^{n+1} / \mathbb{R}_+^\times$ and
hence on the sphere $\mathbb{S}^n$ via

$$g.x := J(g, x)^{-1}(c + dx) = \xi^{-1}(J(g, x)^{-1} g(\xi(x))) \tag{7.11}$$

with

$$J(g, x) := a + \langle b, x \rangle = (g.\xi(x))_0. \tag{7.12}$$

Let

$$K := \left\{ \begin{pmatrix} 1 & 0 \\ 0 & d \end{pmatrix} : d \in O_{n+1}(\mathbb{R}) \right\} \cong O_{n+1}(\mathbb{R}) .$$

Then K is a maximal compact subgroup of G acting transitively on the sphere via
the standard action on $\mathbb{R}^{n+1}$, and $\mathbb{S}^n \cong K/M$ for $M := K_{e_0} \simeq O_n(\mathbb{R})$. Note that K
is the stabiliser of e_{-1} in G with respect to the standard linear action.

As a homogeneous space of G, the sphere is G/P, where $P = G_{e_0}$ is the stabilizer
of e_0. We have $P = MAN$, where

$$A = \left\{ a_t = \begin{pmatrix} \cosh(t) & \sinh(t) & 0 \\ \sinh(t) & \cosh(t) & 0 \\ 0 & 0 & I_n \end{pmatrix} : t \in \mathbb{R} \right\} \cong \mathbb{R}$$

and

$$N := \left\{ n_v = \begin{pmatrix} 1 + \frac{\|v\|^2}{2} & -\frac{\|v\|^2}{2} & v^\top \\ \frac{\|v\|^2}{2} & 1 - \frac{\|v\|^2}{2} & v^\top \\ v & -v & I_n \end{pmatrix} : v \in \mathbb{R}^n \right\} \simeq \mathbb{R}^n.$$

We define

$$J_\lambda(g, v) := J(g, v)^{-\lambda - \frac{n}{2}}, \qquad Q(u, v) := 1 - \langle u, v \rangle \quad \text{and} \quad Q_\lambda(u, v) := Q(u, v)^{\lambda - \frac{n}{2}}.$$

Part (a) of the following lemma is [vD09, Proposition 7.5.8]. It also follows from
[NÓ14, Rem. 5.2] by the transformation formula for integrals, and the remainder is
obtained by direct calculation.

Lemma 7.3.1 *Let $g, g_1, g_2 \in G$ and $u, v \in \mathbb{S}^n$. For the K-invariant probability measure $\mu_{\mathbb{S}^n}$ on $\mathbb{S}^n$, we have*

(a) $\int_{\mathbb{S}^n} \varphi(g.v) J_{\frac{n}{2}}(g, v) \, d\mu_{\mathbb{S}^n}(v) = \int_{\mathbb{S}^n} \varphi(v) \, d\mu_{\mathbb{S}^n}(v)$ *for* $\varphi \in L^1(\mathbb{S}^n)$.
(b) $J_\lambda(g_1 g_2, v) = J_\lambda(g_1, g_2.v) J_\lambda(g_2, v)$.
(c) $Q_\lambda(u, v) = J_{-\lambda}(g, u) Q_\lambda(g.u, g.v) J_{-\lambda}(g, v)$.

Definition 7.3.2 For every $\lambda \in \mathbb{C}$, we obtain a representation of G on $C^\infty(\mathbb{S}^n)$ by

$$(U_g^\lambda \varphi)(v) = J_\lambda(g^{-1}, v) \varphi(g^{-1}.v). \tag{7.13}$$

We denote by C_λ^∞ the space $C^\infty(\mathbb{S}^n)$ with the G-action given by U^λ. Similarly, $C_\lambda^{-\infty}$ will denote the space of distributions with the contragradient action.

From Lemma 7.3.1 we get:

Lemma 7.3.3 *For $\varphi, \psi \in C^\infty(\mathbb{S}^n)$ and $g \in G$, we have*

$$\langle U_g^{-\bar{\lambda}} \varphi, U_g^\lambda \psi \rangle_{L^2} = \langle \varphi, \psi \rangle_{L^2}. \tag{7.14}$$

(a) *The representation U^λ extends to a unitary representation of G on $L^2(\mathbb{S}^n)$ if and only if $\lambda \in i\mathbb{R}$.*
(b) *The linear map $\psi \mapsto \langle \cdot, \psi \rangle_{L^2}$ defines a linear and G-equivariant map from $C_{-\bar{\lambda}}^\infty$ into $C_\lambda^{-\infty}$.*

The following theorem follows from [vD09, Cor. 7.5.12], which is stated for the space $C(\mathbb{S}^n)$ of continuous functions, but the same argument works for smooth functions.

Theorem 7.3.4 *The representation $(U^\lambda, C^\infty(\mathbb{S}^n))$ is irreducible if $\pm\lambda \notin \frac{n}{2} + \mathbb{N}_0$. In particular, the unitary representation $(U^\lambda, L^2(\mathbb{S}^n))$ is irreducible for $\lambda \in i\mathbb{R}$.*

7.3.2 The Complementary Series

In this section we explain how U^λ can be made unitary for $\lambda \in (-\frac{n}{2}, \frac{n}{2})$. As Lemma 7.3.3 easily implies that $U^\lambda \simeq U^{-\lambda}$ holds for the corresponding unitary representations, we shall assume that $\lambda \in (0, \frac{n}{2})$.

Recall that the tangent space at $u \in \mathbb{S}^n$ is given by $T_u(\mathbb{S}^n) \cong u^\perp$ and that the stabilizer of u in K acts by the natural linear action on $T_u(\mathbb{S}^n)$. We also write

$$S_u(\mathbb{S}^n) := \{w \in T_u(\mathbb{S}^n) : \|w\| = 1\}.$$

The Riemannian exponential map $\mathrm{Exp}_u : T_u(\mathbb{S}^n) \to \mathbb{S}^n$ is given by

$$\mathrm{Exp}_u(v) = \cos(\|v\|)u + \frac{\sin(\|v\|)}{\|v\|}v, \tag{7.15}$$

i.e., for $\|v\| = 1$, the geodesic starting in u in the direction of v is given

$$x_u(t, v) := \gamma_v(t) = \cos(t)u + \sin(t)v. \tag{7.16}$$

The map $(t, v) \mapsto \gamma_v(t)$, $t \in (0, \pi)$, $v \in S_u(\mathbb{S}^n)$ defines the *polar coordinates* on $\mathbb{S}^n$.

For further references we recall the following facts about the Beta and Gamma function. The *Beta function* is defined by

$$B(z, w) := \int_0^1 r^{z-1}(1 - r)^{w-1} dr = \frac{\Gamma(z)\Gamma(w)}{\Gamma(z + w)}, \quad \operatorname{Re} z, \operatorname{Re} w > 0.$$

Lemma 7.3.5 *For* $\operatorname{Re} z, \operatorname{Re} w > 0$, *the following assertions hold:*

(a) $B(z, w) = \int_0^\infty \frac{t^{z-1}}{(1+t)^{z+w}} dt.$

(b) $\sqrt{\pi}\Gamma(2z) = 2^{2z-1}\Gamma(z)\Gamma(z + 1/2).$

(c) $\int_{-1}^1 (1 - r)^{z-1}(1 - r^2)^{w-1} dr = 2^{2w+z-2} B(w, w + z - 1).$

(d) *The euclidean surface measure of the sphere is* $\operatorname{Vol}(\mathbb{S}^{n-1}) = 2\frac{\pi^{n/2}}{\Gamma(n/2)}.$

(e) *For* $\operatorname{Re} \sigma > -n$ *and* $\operatorname{Re} \mu > $ n *we have*

$$\int_{\mathbb{R}^n} (1 + \|y\|^2)^{-\mu}\|y\|^\sigma dy = \pi^{n/2}\frac{\Gamma((\sigma + n)/2)\Gamma(\mu - (\sigma + n)/2)}{\Gamma(n/2)\Gamma(\mu)}.$$

Proof (a) follows with $r = \frac{t}{1+t}$ and (b) can be found in [WW63, Sect. 12.15]. Formula (c) follows from (b) by the substitution $u = (1 + r)/2$, (d) is [Fa08, Sect. 9.1], and (e) follows from (a) and (d) by using polar coordinates and substituting $u = r^2$.   □

Lemma 7.3.6 *For* $c_n := \frac{\Gamma(\frac{n+1}{2})}{\sqrt{\pi}\Gamma(\frac{n}{2})}$, *we have*

$$\int_{\mathbb{S}^n} \varphi(u) d\mu_{\mathbb{S}^n}(u) = c_n \int_0^\pi \int_{S_u} \varphi(x_u(t, v)) \sin^{n-1}(t) \, d\mu_{\mathbb{S}^{n-1}}(v) \, dt \quad for \quad \varphi \in L^1(\mathbb{S}^n).$$

If φ *is* K_u-invariant, then $\tilde{\varphi}(\cos t) := \varphi(x_u(t, v))$ *is independent of* $v \in S_u$ *and*

$$\int_{\mathbb{S}^n} \varphi \, d\mu_{\mathbb{S}^n} = c_n \int_{-1}^1 \tilde{\varphi}(r)(1 - r^2)^{\frac{n}{2}-1} dr. \tag{7.17}$$

Proof See [Fa08, Proposition 9.1.2]. The value of the constant follows from Lemma 7.3.5(c) by taking $\lambda = n/2$ and $\varphi = 1$.   □

Lemma 7.3.7 *For* $\lambda \in \mathbb{R}$, *the kernel* Q_λ *is integrable as a function of one or two variables if and only if* $\lambda > 0$. *In that case we have for all* $z \in \mathbb{S}^n$:

$$\int_{\mathbb{S}^n} Q_\lambda(z, y) d\mu_{\mathbb{S}^n}(y) = \frac{2^{\lambda+\frac{n}{2}-1}\Gamma(\frac{n+1}{2})\Gamma(\lambda)}{\sqrt{\pi}\Gamma(\lambda + \frac{n}{2})} =: d_{\lambda,n}$$

Proof As Q_λ and the function $\int_{\mathbb{S}^n} Q_\lambda(\cdot, y) d\sigma(y)$ are K-invariant, we have

$$\int_{\mathbb{S}^n} Q_\lambda(z, y) d\mu_{\mathbb{S}^n}(y) = \int_{\mathbb{S}^n} Q(e_0, y) d\mu_{\mathbb{S}^n}(y) = \int_{\mathbb{S}^n}\int_{\mathbb{S}^n} Q_\lambda(x, y) d\mu_{\mathbb{S}^n}(y) d\mu_{\mathbb{S}^n}(x).$$

The function $Q_\lambda(e_0, \cdot)$ is invariant under K. Lemmas 7.3.6 and 7.3.5(c) imply that

$$\int_{\mathbb{S}^n} Q_\lambda(e_0, y) d\mu_{\mathbb{S}^n}(y) = c_n \int_{-1}^{1} (1-r)^{\lambda - \frac{n}{2}} (1-r^2)^{\frac{n}{2}-1} dr = \frac{2^{\lambda + \frac{n}{2}-1} \Gamma(\frac{n+1}{2}) \Gamma(\lambda)}{\sqrt{\pi} \Gamma(\lambda + \frac{n}{2})}.$$

Clearly the integral is finite if and only if Re $\lambda > 0$.   □

For Re $\lambda > 0$, define

$$(A_\lambda \varphi)(x) := \frac{1}{d_{\lambda,n}} \int_{\mathbb{S}^n} Q_\lambda(x, y) \varphi(y) d\mu_{\mathbb{S}^n}(y) \quad \text{for} \quad \varphi \in C^\infty(\mathbb{S}^n).$$

Then $\langle \varphi, A_\lambda \psi \rangle = \langle A_{\overline{\lambda}} \varphi, \psi \rangle$ and $A_\lambda 1 = 1$. In particular, if λ is real then $\langle \varphi, A_\lambda \psi \rangle = \langle A_\lambda \varphi, \psi \rangle$. Furthermore $A_\lambda : L^2(\mathbb{S}^n) \to L^2(\mathbb{S}^n)$ is bounded if Re $\lambda > \frac{n}{2}$ because in this case the kernel Q_λ is continuous and hence in L^2.

Theorem 7.3.8 *Let* $\varphi \in C^\infty(\mathbb{S}^n)$. *Then the following assertions hold:*

(a) $A_\lambda U_g^\lambda \varphi = U_g^{-\lambda} A_\lambda \varphi$ *for* $g \in G$. *In particular, the form*

$$\langle \varphi, \psi \rangle_\lambda := \langle \varphi, A_\lambda \psi \rangle_{L^2}$$

 is U_G^λ-*invariant if* $\lambda > 0$.
(b) *The map* $\{\lambda \in \mathbb{C} : \text{Re } \lambda > 0\} \to L^2(\mathbb{S}^n)$, $\lambda \mapsto A_\lambda \varphi$, *is holomorphic and has a meromorphic extension to all of* $\mathbb{C}$. *Furthermore, the intertwining relation in* (a) *holds then for almost all* λ.

Proof The first part of (a) follows from Lemma 7.3.1(a, c) and the second part is a consequence of Lemma 7.3.3. For (b), we refer to [VW90, Theorem 1.5] or [vD09, Theorem 9.2.12]. A more direct argument can be based on $\Delta r^{-\lambda} = \lambda(\lambda + 1) r^{-\lambda - 2}$ on $\mathbb{R}^n$.   □

We will now determine those $\lambda > 0$ for which the form $\langle \cdot, \cdot \rangle_\lambda$ is positive semidefinite. For that it is easier to work with the realization in a space of functions on $\mathbb{R}^n$. For that we recall the stereographic projection

$$s : \mathbb{R}^n \to \mathbb{S}^n \setminus \{-e_0\}, \quad x \mapsto \left(\frac{1 - \|x\|^2}{1 + \|x\|^2}, \frac{2x}{1 + \|x\|^2} \right) = n_x^\top . e_0 \qquad (7.18)$$

with inverse $s^{-1}(y_0, \mathbf{y}) = \frac{1}{1+y_0} \mathbf{y}$.

Lemma 7.3.9 *For* $\varphi \in C^\infty(\mathbb{S}^n)$ *put* $\varphi_\lambda(x) := \varphi(s(x))(1 + \|x\|^2)^{-\lambda - \frac{n}{2}}$. *For the positive constants*

$$a_n := \frac{2^{n-1}\Gamma(\frac{n+1}{2})}{\pi^{\frac{n+1}{2}}} \quad and \quad b_\lambda = \frac{\Gamma(\lambda + n/2)}{\pi^{\frac{n}{2}}\Gamma(\lambda)},$$

we then have

(a) $\displaystyle\int_{\mathbb{S}^n} \varphi(v)\, d\mu_{\mathbb{S}^n}(v) = a_n \int_{\mathbb{R}^n} \varphi(s(x))(1 + \|x\|^2)^{-n}\, dx$ *for* $\varphi \in L^1(\mathbb{S}^n)$.

(b) $Q_\lambda(s(x), s(y)) = 2^{\lambda - \frac{n}{2}}(1 + \|x\|^2)^{-\lambda + \frac{n}{2}}(1 + \|y\|^2)^{-\lambda + \frac{n}{2}}\|x - y\|^{2\lambda - n}$.

(c) $(A_\lambda\varphi)(s(x)) = (1 + \|x\|^2)^{-\lambda + n/2}b_\lambda \displaystyle\int_{\mathbb{R}^n} \varphi_\lambda(y)\|x - y\|^{2\lambda - n}\, dy$

(d) $\langle \varphi, \psi \rangle_\lambda = a_n b_\lambda \displaystyle\int_{\mathbb{R}^n \times \mathbb{R}^n} \overline{\varphi_\lambda(x)}\psi_\lambda(y)\|x - y\|^{2\lambda - n}\, dx\, dy$ *for* $\varphi, \psi \in C^\infty(\mathbb{S}^n)$.

Proof Up to a constant (a) follows from [Fa08, Example 9.1]. The exact value of the constant can then be evaluated using Lemma 7.3.5. Parts (b) and (c) follow from (a) and Lemma 7.3.1(b,c). Finally (d) follows from (c). $\qquad\square$

Proposition 7.3.10 *The function* $x \mapsto \|x\|^{-s}$ *is locally integrable on* $\mathbb{R}^n$ *if and only if* $s < n$. *The corresponding distribution is positive definite if and only if* $0 \leq s < n$.

Proof This follows by using polar coordinates and the fact that

$$\mathcal{F}(r^{-s}) = \pi^{s - n/2}\frac{\Gamma((n - s)/2)}{\Gamma(s/2)}r^{s-n}$$

(see [Sch73, Example 5, VII.7.13]). The right hand side is positive for $0 < s < n$. The case $s = 0$ is obvious. $\qquad\square$

As a consequence we get the following theorem, up to the non-degeneracy of the form:

Theorem 7.3.11 *For* $\lambda \geq 0$ *the form* $\langle \cdot, \cdot \rangle_\lambda$ *is positive semi-definite on* $C^\infty(\mathbb{S}^n)$ *if and only if* $0 < \lambda \leq \frac{n}{2}$. *Let* $\mathscr{E}_\lambda$ *denote the corresponding Hilbert space. For* $\lambda = \frac{n}{2}$ *this space is one-dimensional and for* $0 < \lambda < \frac{n}{2}$ *the form is non-degenerate. We thus obtain irreducible unitary representations* $(U^\lambda, \mathscr{E}_\lambda)_{0 < \lambda \leq \frac{n}{2}}$, *where* $(U^{n/2}, \mathscr{E}_{n/2})$ *is trivial.*

Proof For $0 < \lambda < \frac{n}{2}$ the non-degeneracy of the kernel on $C^\infty(\mathbb{S}^n)$ follows from Theorem 7.3.4 which asserts that the representation U^λ on $C^\infty(\mathbb{S}^n)$ is irreducible. As the space of null-vectors is invariant and proper, it is zero. $\qquad\square$

Definition 7.3.12 The representations $(U^\lambda, \mathscr{E}_\lambda)$, $0 < \lambda < \frac{n}{2}$, are called the *complementary series representations* of G.

To unify notation we put $\mathscr{E}_\lambda = L^2(\mathbb{S}^n)$ for $\lambda \in i\mathbb{R}$ (cf. Lemma 7.3.3). The proof of the following can be found in [vD09, p. 119]. We shall encounter this theorem again in the next section.

Theorem 7.3.13 *The irreducible unitary representations $(U, \mathcal{E})$ of G which are spherical in the sense that $\mathcal{E}^K \neq \{0\}$ are exactly the representations $(U^\lambda, \mathcal{E}_\lambda)$ with $\lambda \in i\mathbb{R} \cup (0, \frac{n}{2}]$. In these cases $\mathcal{E}^K = \mathbb{C}1$ is one dimensional.*

The function $\varphi_\lambda(g) = \langle 1, U_g^\lambda 1 \rangle_\lambda$ is K-biinvariant. It is called the *spherical function with spectral parameter* λ.

7.3.3 H-Invariant Distribution Vectors

On $G = \mathrm{O}_{1,n+1}(\mathbb{R})^\uparrow$ we define an involution $\tau : G \to G$ by $\tau(g) := r_0 g r_0$, where r_j is the orthogonal reflection in $e_j^\perp$. Then, with respect to the linear action on $\mathbb{R}^{n+2}$,

$$H := G_{e_0} = \left\{ \begin{pmatrix} a & 0 & b^\top \\ 0 & 1 & 0 \\ c & 0 & d \end{pmatrix} \in \mathrm{O}_{1,n+1}(\mathbb{R})^\uparrow : a \in \mathbb{R}, b, c \in \mathbb{R}^{n-1}, d \in M_{n-1}(\mathbb{R}) \right\} \subsetneq G^\tau$$

is a subgroup isomorphic to $\mathrm{O}_{1,n}(\mathbb{R})^\uparrow$. The relation $r_0 \xi(v) = \xi(r_0 x)$ implies that $r_0(x.v) = \tau(x).r_0(v)$. Here we have also viewed r_0 as a reflection in $\mathbb{R}^{n+1}$ via restriction.

In $\mathbb{S}^n$, the subgroup H has two open orbits

$$H.(\pm e_0) = \mathbb{S}_\pm^n = \{(x_0, \mathbf{x}) : \pm x_0 > 0\}$$

and the closed orbit $H.e_n = \{(0, \mathbf{x}) : \mathbf{x} \in \mathbb{S}^{n-1}\} \cong \mathbb{S}^{n-1}$.

Considering the standard linear action of G on $\mathbb{R}^{n+2}$, we note that $G.e_0 \cong G/H$ is the $(n+1)$-dimensional de Sitter space

$$\mathrm{dS}^{n+1} := \{(x_{-1}, x_0, \mathbf{x}) : x_{-1}^2 - x_0^2 - \|\mathbf{x}\|^2 = -1\}.$$

Both G and H are unimodular. Hence, there exists a G-invariant measure on dS^{n+1}.
Define

$$p_\lambda^\pm(x) := \frac{[\xi(x), \mp e_0]^{\bar{\lambda} - \frac{n}{2}}}{\Gamma((\bar{\lambda} - \frac{n}{2} + 1)/2)} \chi_{\mathbb{S}_\pm^n}(x) = \frac{(\pm x_0)^{\bar{\lambda} - \frac{n}{2}}}{\Gamma((\bar{\lambda} - \frac{n}{2} + 1)/2)} \chi_{\mathbb{S}_\pm^n}(x), \quad x \in \mathbb{S}^n$$

and let

$$p_\lambda := p_\lambda^+ + p_\lambda^-. \tag{7.19}$$

For $\operatorname{Re} \lambda > n/2$ the functions $p_\lambda^\pm$ and p_λ are continuous and hence integrable on $\mathbb{S}^n$. We define a distribution $\iota_{-\infty}(p_\lambda) := \tilde{\eta}_\lambda \in C_\lambda^{-\infty}$ by

$$\tilde{\eta}_\lambda(\varphi) := \int_{\mathbb{S}^n} \overline{\varphi(x)} p_\lambda(x) \, d\mu_{\mathbb{S}^n}(x) = \langle \varphi, p_\lambda \rangle_{L^2} \quad \text{for} \quad \operatorname{Re} \lambda > \frac{n}{2}.$$

Then $\lambda \mapsto \tilde{\eta}_\lambda(\varphi)$ is antiholomorphic on $\{\lambda \in \mathbb{C} : \operatorname{Re}(\lambda) > \frac{n}{2}\}$. We define $\tilde{\eta}_\lambda^\pm$ in the same way.

Theorem 7.3.14 *The families of distributions $\tilde{\eta}_\lambda$, $\tilde{\eta}_\lambda^\pm$ are antiholomorphic for $\operatorname{Re} \lambda > \frac{n}{2}$ and have an antiholomorphic extension to $\mathbb{C}$. The distributions $\tilde{\eta}_\lambda$, $\tilde{\eta}_\lambda^\pm$ are H-invariant and for almost all λ we have $(C_\lambda^{-\infty})^H = \mathbb{C}p_\lambda^+ + \mathbb{C}p_\lambda^-$.*

Proof For the analytic continuation we refer to [vD09, Proposition 9.2.9]. It is clear that p_λ is r_0-invariant. A simple calculation shows that $J_{-\bar{x}}(h, x)^{-1} p_\lambda(h.x) = p_\lambda(x)$ which implies that $U_h^{\bar{\lambda}} p_\lambda = p_\lambda$ for $h \in H$. Hence $\tilde{\eta}_\lambda$ is H-invariant for $\operatorname{Re} \lambda < -\frac{n}{2}$. The uniqueness of analytic extension then implies the assertion for all λ. The last statement can be found in [vB88, Theorem 5.10]. ☐

For Proposition 7.1.6 we adjust the definition so that $\eta_\lambda := \tilde{\eta}_\lambda$ for $\lambda \in i\mathbb{R}$ and $\eta_\lambda(\varphi) = \tilde{\eta}_{-\lambda}(A_\lambda \varphi)$ for $\lambda \in (0, \frac{n}{2})$. Then η_λ is still invariant under H and τ. Furthermore, as U^λ is irreducible and $\eta_\lambda \neq 0$, it follows that η_λ is cyclic. Hence Theorem 7.1.7 gives:

Theorem 7.3.15 *The unitary representation $(U^\lambda)_{\lambda \in i\mathbb{R} \cup (0, \frac{n}{2})}$ can be realized in a Hilbert space of distributions on de Sitter space $\mathrm{dS}^{n+1} \cong G/H$.*

7.4 Reflection Positivity

We now turn to reflection positivity, as it manifests itself for spherical representations of the Lorentz group.

7.4.1 Reflection Positivity for the Conformal Group

In this section we discuss the reflection positivity of the representation $(U^\lambda, \mathscr{E}_\lambda)$ of $G = O_{1,n+1}(\mathbb{R})^\uparrow$ for $\lambda \in (0, \frac{n}{2})$. We consider again the involutions τ and r_0. Define $\theta : \mathscr{E}_\lambda \to \mathscr{E}_\lambda$ by $\theta(\varphi) := \varphi \circ r_0$. Then $\theta p_\lambda = p_\lambda$, $\theta(U_g^\lambda \varphi) = U_{\tau(g)}^\lambda \theta \varphi$ and $A_\lambda \theta = \theta A_\lambda$. In particular

$$\langle \theta \varphi, \psi \rangle_\lambda = \int_{\mathbb{S}^n \times \mathbb{S}^n} \overline{\varphi(x)} \psi(y) Q_\lambda(r_0 x, y) \, d\mu_{\mathbb{S}^n}(x) \, d\mu_{\mathbb{S}^n}(y). \tag{7.20}$$

We let $\mathscr{E}_{\lambda+}$ be the space generated by the functions supported by the half sphere $\mathbb{S}_+^n$. For the positivity of the twisted inner product on $\mathscr{E}_{\lambda+}$ we switch to the realization of $(U^\lambda, \mathscr{E}_\lambda)$ as acting on functions on $\mathbb{R}^n$ via the stereographic projection s from (7.18).

Lemma 7.4.1 *Let $R_\lambda(x, y) := (1 - \langle x, y \rangle + \|x\|^2 \|y\|^2)^{\lambda - \frac{n}{2}}$ and define $\sigma : \mathbb{R}^n \to \mathbb{R}^n$ by $\sigma(x) = \frac{x}{\|x\|^2}$. Then the following holds:*

(a) *The stereographic projection $s : B_1(0) \to \mathbb{S}_+^n$ from (7.18) is a diffeomorphism from the open unit ball $B_1(0) \subseteq \mathbb{R}^n$ onto $\mathbb{S}_+^n$. In particular, $\operatorname{supp} \varphi \subset \mathbb{S}_+^n$ if and only if $\operatorname{supp}(\varphi \circ s) \subset B_1(0)$.*

(b) $\sigma = s^{-1} \circ r_0 \circ s$.

(c) $\|\sigma(x) - y\|^2 = \|x\|^{-2}(1 - 2\langle x, y\rangle + \|x\|^2 \|y\|^2)$.

(d) $\langle \theta\varphi, \psi\rangle_\lambda = e_\lambda \int_{B_1(0) \times B_1(0)} \overline{\varphi_\lambda(x)}\psi_\lambda(y) R_\lambda(x, y)\, dx\, dy$ *for $\varphi, \psi \in C_c^\infty(\mathbb{S}_+^n)$.*

Proof (a) follows from $1 - \|x\|^2 > 0$ if and only if $\|x\| < 1$ and (b) and (c) are simple calculations and then (d) follows from (c), Lemma 7.3.9 and (7.20).   □

Theorem 7.4.2 ([NÓ14, Proposition 6.2]) *Let $n \geq 1$. The kernel R_λ is positive definite on $B_1(0)$ if and only if $\lambda = \frac{n}{2}$ or $\lambda \leq \min\{\frac{n}{2}, 1\}$.*

The group $H = \mathrm{O}_{1,n}(\mathbb{R})^\uparrow$ maps $\mathbb{S}_+^n$ into itself, so that $U_H^\lambda \mathscr{E}_{\lambda,+} = \mathscr{E}_{\lambda,+}$. Furthermore $dU^\lambda(\mathfrak{g})C^\infty(\mathbb{S}_+^n) \subseteq C^\infty(\mathbb{S}_+^n)$. The subsemigroup $S := \{s \in G : s.\overline{\mathbb{S}_+^n} \subset \mathbb{S}_+^n\}$ is open and $\sharp$-invariant with $e \in \overline{S}$. Combining Theorem 7.4.2 with Theorem 7.3.11, we obtain:

Theorem 7.4.3 *For $n \geq 2$, the following assertions hold:*

(a) *The subspace space $\mathscr{E}_{\lambda,+}$ is U_S^λ-invariant for all $\lambda \in i\mathbb{R} \cup (0, n/2)$.*

(b) *For $\lambda \in i\mathbb{R}$ we have $\mathscr{E}_{\lambda,+} \perp \theta\mathscr{E}_{\lambda,+}$, so that $(\mathscr{E}_\lambda, \mathscr{E}_{\lambda,+}, \theta)$ is reflection positive with $\widehat{\mathscr{E}}_\lambda = \{0\}$.*

(c) *For $0 < \lambda \leq \frac{n}{2}$, the triple $(\mathscr{E}_\lambda, \mathscr{E}_{\lambda,+}, \theta)$ and the distribution vector η_λ are reflection positive with respect to (G, S, τ) if and only if $\lambda \leq 1$. In this case $\widehat{\mathscr{E}}_\lambda$ is infinite dimensional except for $n = 2$ and $\lambda = 1$, where $\widehat{\mathscr{E}}_\lambda$ is one dimensional.*

Remark 7.4.4 (a) The domain where R_λ is positive definite includes the half-line, $\lambda \leq \min\{1, \frac{n}{2}\}$. On this half line we always have an G-invariant hermitian form on C_λ^∞ which is positive definite only for $\lambda \geq 0$. This leads to the situation where we have a Fréchet space with a G-invariant hermitian form which is not positive definite, but the induced form on $\mathscr{E}_{\lambda,+}$ is positive leading to a OS-quantization for a non-unitary representation of G. For detailed discussion see [FÓØ18, JÓl98, JÓl00, Ól00].

(b) The group G^c is the universal covering of the group $\mathrm{SO}_{2,n}(\mathbb{R})_0$. It acts transitively on the Lie ball

$$\mathscr{D} := \{z = \xi + i\eta \in \mathbb{C}^n : \xi^2 + \eta^2 + 2\sqrt{\xi^2\eta^2 - (\xi\eta)^2} < 1\}$$

The stabilizer of $0 \in \mathscr{D}$ is the universal cover K^c of $S(\mathrm{O}_2(\mathbb{R}) \times \mathrm{O}_n(\mathbb{R}))_0$ and $\mathscr{D} \cong G^c/K^c$.

The real ball $B_1(0)$ is a totally real submanifold in $\mathscr{D}$. Furthermore

$$R_\lambda(z, w) := (1 - z\overline{w} + z^2\overline{w}^2)^{\lambda - \frac{n}{2}}, \tag{7.21}$$

where $st = \sum s_j t_j$ and $s^2 = ss$, is well-defined, holomorphic in the first variable, antiholomorphic in the second variable and $R_\lambda(z, w) = \overline{R_\lambda(w, z)}$. Thus R_λ is a hermitian kernel on $\mathscr{D}$ and positive definite if $\lambda = \frac{n}{2}$ or $\lambda \leq \min\{1, \frac{n}{2}\}$. The representation

U^{λ_C}, which exists by Theorem 6.2.3, is a *negative energy representation* of G^c (i.e., a *highest weight representation*). We refer to [FK94, Chap. XIII] and [Ne00] for detailed discussion of such representations.

7.4.2 Resolvents of the Laplacian on the Sphere

Now we continue the example of the sphere by specializing the construction from Sect. 2.5 based on resolvents $(m^2 - \Delta)^{-1}$, $m > 0$, of the Laplacian on $\mathbb{S}^n$ with the involution r_0. In this section the starting group will be $O_{n+1}(\mathbb{R})$ acting on the sphere $\mathbb{S}^n$, whereas $O_{1,n}(\mathbb{R})^\uparrow$ will play the role of the dual group. We therefore change our notation a little and let $G = O_{n+1}(\mathbb{R})$ and $K = G_{e_0} \cong O_n(\mathbb{R})$. Accordingly, reflection positivity leads to unitary representations of G^c depending on the parameter m. Accordingly, in the discussion about the representations $(U^\lambda, \mathscr{E}^\lambda)$ the n in Sect. 7.3 will change to $n - 1$.

We start with some general simple facts.

Lemma 7.4.5 *Let τ be a dissecting reflection on the connected complete Riemannian manifold M and $m > 0$. Let $C = (m^2 - \Delta_M)^{-1}$ and θ be as in Theorem 2.5.1. Let $\Theta : M \to M$ be an isometric diffeomorphism. Then the following assertions hold:*

(a) *$C \circ \Theta_* = \Theta_* \circ C$.*
(b) *Let D be the reflection positive distribution defined by $D(\varphi \otimes \psi) = \langle \varphi, C\overline{\psi} \rangle_{L^2(M)}$. Then $D((\Theta_*\varphi) \otimes \psi) = D(\varphi \otimes (\Theta_*^{-1}\psi))$.*

Note that $(m^2 - \Delta_M)D(x, y) = \delta_M(x, y)$, where the distribution δ_M on $M \times M$ is given by $\delta_M(\varphi) = \int_M \varphi(x, x) \, dV_M(x)$, where V_M denotes the volume measure on M. This implies in particular

$$(m^2 - \Delta_M)_x D = (m^2 - \Delta_M)_y D = (m^2 - \Delta_M)_x (m^2 - \Delta_M)_y D = 0$$

off the diagonal in $M \times M$ and that $(\Delta_M - m)_x (\Delta_M - m)_y$ is an elliptic differential operator on $M \times M$ we have:

Lemma 7.4.6 *On the open submanifold $(M \times M) \setminus \mathrm{diag}(M)$, the distribution D is represented by an analytic function Φ, which is invariant under the isometry group $\mathrm{Isom}(M)$.*

Define $C^\tau := C \circ \tau_*$. Then, by the above lemma, there exists an analytic function Ψ on $M_+ \times M_+$ such that

$$(C^\tau \varphi)(x) = \int_{M_+} \Psi(x, y)\varphi(y)dV_M(y) \quad \text{for} \quad x \in M_+.$$

As $(m^2 - \Delta_M)C^\tau\varphi = 0$ for $\varphi \in C_c^\infty(M_+)$ it follows that $C^\tau\varphi$ is analytic on M_+ and $\Delta_M(C^\tau\varphi|_{M_+}) = m^2 C^\tau\varphi|_{M_+}$.

Since cyclic one-parameter semigroups of contractions are given by multiplication with functions on L^2-spaces, we take a closer look at this special situation. We give here one example, a second example is discussed in the following subsection.

After these general remarks, we now specialize to

$$M = \mathbb{S}^n \supset M_+ = \mathbb{S}^n_+ = \{x \in \mathbb{S}^n : x_0 > 0\}.$$

To work with the exponential function we introduce the analytic functions

$$C(z) := \sum_{k=0}^{\infty} \frac{(-1)^k}{(2k)!} z^k \quad \text{and} \quad S(z) := \sum_{k=0}^{\infty} \frac{(-1)^k}{(2k+1)!} z^k \tag{7.22}$$

which satisfy $\cos z = C(z^2)$ and $\frac{\sin z}{z} = S(z^2)$ for $z \in \mathbb{C}^\times$. We thus obtain as in (7.15) that

$$\text{Exp}_u(v) = C(v^2) \cdot u + S(v^2) \cdot v, \quad u \in \mathbb{S}^n, v \in S_u. \tag{7.23}$$

The complex sphere

$$\mathbb{S}^n_{\mathbb{C}} = \{u \in \mathbb{C}^{n+1} : u^2 = 1\} = O_{n+1}(\mathbb{C}).e_0 \cong O_{n+1}(\mathbb{C})/O_n(\mathbb{C})$$

also is a symmetric space (in the category of complex manifolds) with respect to the reflections $s_x(y) := y - 2(xy)x$, for $x, y \in \mathbb{S}^n_{\mathbb{C}}$ and the corresponding exponential map is

$$\text{Exp}_u(v) = C(v^2) \cdot u + S(v^2) \cdot v \quad \text{for} \quad u \in \mathbb{S}^n_{\mathbb{C}}, v \in T_u(\mathbb{S}^n_{\mathbb{C}}). \tag{7.24}$$

Definition 7.4.7 Let $\iota(x) = (x_0, ix)$ and $V := \iota\mathbb{R}^{n+1} = \mathbb{R}e_0 \oplus i\mathbb{R}^n$. Define

$$[\iota x, \iota y]_V := \iota x \cdot \iota y = x_0 y_0 - \mathbf{xy},$$

and note that $[g.u, g.v]_V = [u, v]_V$ for $u, v \in V$ and all elements $g \in G^c := \iota O_{1,n}(\mathbb{R})^\uparrow \iota$.

On $\mathbb{C}^{n+1}$ we consider the conjugations

$$\sigma_{\mathbb{R}}(z_0, \dots, z_n) := (\overline{z_0}, \dots, \overline{z_n}) \quad \text{and} \quad \sigma_V(z_0, \dots, z_n) := (\overline{z_0}, -\overline{z_1}, \dots, -\overline{z_n})$$

with respect to the real subspaces $\mathbb{R}^{n+1}$ and V, respectively. The conjugations $\sigma_{\mathbb{R}}$ and σ_V commute and the holomorphic involution $\sigma_{\mathbb{R}}\sigma_V$ is $-r_0$. The involution σ_V commutes with G^c, but $\sigma_{\mathbb{R}}$ does not, and $\sigma_{\mathbb{R}} g \sigma_{\mathbb{R}} = r_0 g r_0 = \tau(g)$ is the involution on G^c whose fixed point group is $K = G^c_{e_0} \cong O_n(\mathbb{R})$.

We also note that $[x, y]_V = xy = \sum_j x_j y_j$ for $x, y \in \mathbb{C}^{n+1}$ is the unique complex bilinear extension of $[\cdot, \cdot]_V$ to $V + iV = \mathbb{C}^{n+1}$. This notation underlines the

Lorentzian nature of the situation rather than the euclidean one. We also consider the following sets:

- $V_+ := \{v \in V : [v, v] > 0, v_0 > 0\} = \iota(\mathbb{R}^{1,n}_+)$, the forward light cone in V;
- $\mathbb{H}^n_V := \iota\mathbb{H}^n = \mathbb{S}^n_{\mathbb{C}} \cap V_+ = G^c.e_0 \cong G^c/K$, the hyperboloid of one sheet in V;
- $\mathbb{S}^n_{+,\mathbb{C}} := \{z \in \mathbb{S}^n_{\mathbb{C}} : \operatorname{Re} z_0 > 0\}$;
- the tube domain $T_{V_+} := iV + V_+ \cong SO_{2,n+1}(\mathbb{R})/S(O_2(\mathbb{R}) \times O_{n+1}(\mathbb{R}))$; and
- $\Xi := G^c.\mathbb{S}^n_+ \subset \mathbb{S}^n_{+,\mathbb{C}}$.

The domain Ξ is called the *crown* of $\mathbb{H}^n_V$. Note that $G^c.V = V$, $G^c.T_{V_+} = T_{V_+}$ and $G^c.\mathbb{S}^n_{\mathbb{C}} = \mathbb{S}^n_{\mathbb{C}}$.

Proposition 7.4.8 *The following assertions hold:*

(a) $T_{V_+} \cap \mathbb{R}^{n+1} = \mathbb{R}^{n+1}_+$ *and* $\mathbb{S}^n_+ = T_{V_+} \cap \mathbb{S}^n$.

(b) $\Xi = T_{V_+} \cap \mathbb{S}^n_{+,\mathbb{C}} = T_{V_+} \cap \mathbb{S}^n_{\mathbb{C}}$.

(c) *We have* $\sigma_V \Xi = \sigma_{\mathbb{R}} \Xi = \Xi$ *and* $\Xi^{\sigma_V} = \Xi \cap V = \mathbb{H}^n_V$ *and* $\Xi^{\sigma_{\mathbb{R}}} = \Xi \cap \mathbb{R}^{n+1} = \mathbb{S}^n_+$.

(d) $\mathbb{C}_\Xi := \{z \in \mathbb{C} : (\exists u, v \in \Xi)\, z = [u, v]_V\} = \mathbb{C} \setminus (-\infty, -1]$.

Proof (a) This follows from $z = u + iv \in T_{V_+} \cap \mathbb{R}^{n+1}$ if and only if $u = re_0$ with $r > 0$ and $iv = (0, \mathbf{v})$ with $\mathbf{v} \in \mathbb{R}^n$.

(b) By (i) we have $\Xi = G^c.\mathbb{S}^n_+ = G^c.(T_{V_+} \cap \mathbb{S}^n) \subseteq T_{V_+} \cap \mathbb{S}^n_{\mathbb{C}}$. Let $z = u + iv \in T_{V_+} \cap \mathbb{S}^n_{\mathbb{C}}$. Then $u_0 > 0$ and, as G^c acts transitively on all level sets $[u, u] = r > 0$ in V_+, we may assume that $u = re_0$ with $r > 0$. Thus $z = (r + iv_0, \mathbf{v})$ with $v_0 \in \mathbb{R}$ and $\mathbf{v} \in \mathbb{R}^n$. As $z \in \mathbb{S}^n_{\mathbb{C}}$, we have $1 = zz = r^2 - v_0^2 + 2irv_0 + \mathbf{v}^2$. Hence $v_0 = 0$ and this implies that $z \in \mathbb{S}^n_+ \subset \Xi$. Finally, we note that, if $z \in T_{V_+}$, then $\operatorname{Re} z_0 > 0$, hence $T_{V_+} \cap \mathbb{S}^n_{\mathbb{C}} = T_{V_+} \cap \mathbb{S}^n_{+,\mathbb{C}}$.

(c) follows from (a) and (b).

(d) follows from a lengthy, but elementary, calculation; see [NÓ18, Lemma 3.8] for details. $\square$

We point out the following two simple consequences of Proposition 7.4.8

Corollary 7.4.9 *The subset Ξ is an open complex submanifold of $\mathbb{S}^n_{\mathbb{C}}$ on which the group G^c acts by holomorphic maps. The fixed point set fixed point sets $\mathbb{S}^n_+ = \Xi^{\sigma_{\mathbb{R}}}$ and $\mathbb{H}^n_V = \Xi^{\sigma_V}$ of the antiholomorphic involutions $\sigma_{\mathbb{R}}$ and σ_V are totally real submanifolds of Ξ.*

7.4.3 The Distribution Kernel of $(m^2 - \Delta)^{-1}$

In this section we use polar-coordinates $x_{e_0}(t, u) = \cos(t)e_0 + \sin(t)u$. If $\varphi : \mathbb{S}^n \to \mathbb{C}$ is K-invariant, then it is determined by the values on $x_{e_0}(t, e_n)$, $0 \le t \le \pi$, and we may simply write $\tilde\varphi(t) := \varphi(x_{e_0}(t, e_n))$. Let $\mathbb{S}^n_* := \mathbb{S}^n \setminus \{-e_0\}$ and note that $\mathbb{S}^n_*$ is K-invariant as K fixes $\pm e_0$.

Lemma 7.4.10 ([Fa08, Cor. 9.2.4]) *For $\varphi \in C^2(\mathbb{S}_*^n)^K$ and $t \in (0, \pi)$*

$$(\Delta_{\mathbb{S}^n}\varphi)(x_{e_0}(t, u)) = \tilde{\varphi}''(t) + (n - 1)\cot(t)\tilde{\varphi}'(t) = \frac{1}{\sin^{n-1}(t)}\frac{d}{dt}\left(\sin^{n-1}(t)\frac{d}{dt}\right)\tilde{\varphi}(t).$$

In particular, $\Delta_{\mathbb{S}^n}\varphi = m^2\varphi$ if and only if

$$\tilde{\varphi}''(t) + (n - 1)\cot(t)\tilde{\varphi}'(t) - m^2\tilde{\varphi}(t) = 0. \tag{7.25}$$

We already discussed the case $n = 1$ in Chap. 5 and will therefore assume that $n > 1$. The substitution $s = \sin^2(t/2) = \frac{1-\cos t}{2}, 0 \leq t \leq \pi$, and $\xi(s) := \varphi(x_{e_0}(t, e_n))$ transforms (7.25) into the *hypergeometric differential equation*:

$$s(1 - s)\xi''(s) + \left(\frac{n}{2} - ns\right)\xi'(s) - m^2\xi(s) = 0. \tag{7.26}$$

As ξ does not have a singularity in $s = 0$, this leads to

$$\xi(s) = c \cdot {}_2F_1\left(\frac{n-1}{2} + \lambda, \frac{n-1}{2} - \lambda, \frac{n}{2}; s\right) \tag{7.27}$$

with

$$c = \xi(0) \qquad \text{and} \qquad \lambda := \begin{cases} \sqrt{\left(\frac{n-1}{2}\right)^2 - m^2} & \text{if } m^2 < \left(\frac{n-1}{2}\right)^2 \\ i\sqrt{m^2 - \left(\frac{n-1}{2}\right)^2} & \text{if } m^2 \geq \left(\frac{n-1}{2}\right)^2. \end{cases} \tag{7.28}$$

We recall here the definition of ${}_2F_1$ and refer to [WW63, Sect. 14.2] for more details. For $\alpha \in \mathbb{C}$ and $k \in \mathbb{N}$ let

$$(\alpha)_k := \prod_{j=0}^{k-1}(\alpha + j) = \Gamma(\alpha + k)/\Gamma(\alpha).$$

For $a, b \in \mathbb{C}$ and $c \in \mathbb{C} \setminus -\mathbb{N}_0$, we have

$${}_2F_1(a, b, c; z) = \sum_{k=0}^{\infty} \frac{(a)_k(b)_k}{(c)_k}\frac{z^k}{k!} \qquad \text{for} \quad |z| < 1. \tag{7.29}$$

For $a, b, c > 0$ or $b = \bar{a}, c > 0$, (7.29) implies that ${}_2F_1(a, b; c; z) > 0$ for $0 \leq z < 1$, and ${}_2F_1(a, b; c; 0) = 1$. As ${}_2F_1(a, b; c; z) = {}_2F_1(b, a; c; z)$ and $m > 0$, we obtain the same function if we replace λ by $-\lambda$. In particular,

$${}_2F_1\left(\frac{n-1}{2} + \lambda, \frac{n-1}{2} - \lambda; \frac{n}{2}; x\right) > 0 \quad \text{for} \quad 0 \leq x < 1.$$

We now apply this to the kernel $\Psi_m(u, v)$ corresponding to C^τ. Because of the G-invariance it follows that the function $\psi := \Psi_m(\cdot, e_0) \in C^\infty(\mathbb{S}^n_*)$ is K-invariant. It also satisfies the differential equation $\Delta_{\mathbb{S}^n}\psi = m^2\psi$ on $\mathbb{S}^n_*$ (Sect. 7.4.2). Hence there exists a constant γ such that

$$\psi(x_{e_0}(t, v)) = \gamma \cdot {}_2F_1\left(\frac{n-1}{2} + \lambda, \frac{n-1}{2} - \lambda, \frac{n}{2}; \sin^2(t/2)\right).$$

Theorem 7.4.11 *For every $m > 0$, there exists a constant $\gamma_m > 0$ such that the G^c-invariant kernel Ψ_m, corresponding to C^τ, extends to a hermitian kernel on $\Xi \times \Xi$ is given by*

$$\Psi_m(x, y) = \gamma_m \cdot {}_2F_1\left(\frac{n-1}{2} + \lambda, \frac{n-1}{2} - \lambda; \frac{n}{2}; \frac{1 - [x, \sigma_V(y)]_V}{2}\right) \qquad (7.30)$$

and defines a positive definite hermitian kernel on $\Xi \times \Xi$.

Proof First note that $\sin^2(t/2) = \frac{1}{2}(1 - \cos(t)) = \frac{1}{2}\left(1 - [x_{e_0}(t, v), \sigma_V(e_0)]_V\right)$ for $v \in e_0^\perp$. Hence, by the above discussion:

$$\Psi_m(x, e_0) = c \cdot {}_2F_1\left(\frac{n-1}{2} + \lambda, \frac{n-1}{2} - \lambda, \frac{n}{2}; \frac{1 - [x, \sigma_V(e_0)]_V}{2}\right).$$

The hypergeometric function ${}_2F_1$ has an analytic continuation to $\mathbb{C} \setminus [1, \infty)$, see [WW63, p. 288]. It follows from Proposition 7.4.8(d) that the right hand side, and hence also $\Psi_m(\cdot, e_0)$, has an extension to $\Xi \times \Xi$ given by (7.30). The extension is unique as $\mathbb{S}^n_+$ is a totally real submanifold in Ξ (Proposition 7.4.8(c)). It is holomorphic in the first variable and antiholomorphic in the second variable and $\overline{\Psi_m(x, y)} = \Psi_m(y, x)$ which follows from (7.29) and the form of the parameters. As the kernel C is reflection positive, $\Psi_m(u, v)$ is positive definite on $\mathbb{S}^n_+ \times \mathbb{S}^n_+$. Hence Theorem A.1 in [NÓ14] implies that Ψ_m is positive definite on $\Xi \times \Xi$. In particular $\Psi_m(e_0, e_0) > 0$. It follows that $c > 0$. $\qquad\qquad\qquad\square$

Above we introduced the principal and complementary series representations $(U^\lambda, \mathscr{E}^\lambda)_{\lambda \in i\mathbb{R} \cup (0, \frac{n-1}{2})}$ and just after Theorem 7.3.13 we defined the spherical function $\varphi_\lambda(g)$ with spectral parameter λ. We will use this now for to the group G^c. The following proposition follows from [ÓP13, p. 1158] and [vD09, p. 126]:

Proposition 7.4.12 *For $\psi_m(x) = \frac{1}{c}\Psi_m(x, e_o)$ and $\lambda = \sqrt{\left(\frac{n-1}{2}\right)^2 - m^2}$ as before, we have $\psi_m|_{\mathbb{H}^n} = \varphi_\lambda$. In particular, the spherical function φ_λ extends to a holomorphic function on Ξ and the unitary representation of G^c on the reproducing kernel space $\mathscr{H}_{\Psi_m} \subseteq \mathcal{O}(\Xi)$ with kernel Ψ_m is equivalent to $(U^\lambda, \mathscr{E}^\lambda)$.*

The last statement on analytic continuation in Proposition 7.4.12 is a special case of the general theorem due to Krötz and Stanton [KS04, KS05].

In view of Proposition 7.4.12, we will from now on use the notation φ_λ and Φ_λ for the normalized version of ψ_m and Ψ_m, where λ is defined by (7.28).

Lemma 7.4.13 *Let* $\Psi \neq 0$ *be a hermitian positive definite* G^c-*invariant kernel on* $\Xi \times \Xi$. *If* $\Psi_{e_0}|_{\mathbb{H}^n_V} = 0$, *then* $\Psi = 0$.

Proof As $\mathbb{H}^n_V$ is totally real in Ξ by Corollary 7.4.9, $\Psi_{e_0}|_{\mathbb{H}^n_V} = 0$ implies $\Psi_{e_0} = 0$ on all of Ξ. By the G^c-invariance it follows that $\Psi(z, x) = 0$ for all $z \in \Xi$ and all $x \in \mathbb{H}^n_V$. As $\Psi(z, \cdot)$ is antiholomorphic, it vanishes on all of Ξ.   □

Theorem 7.4.14 *If* Ψ *is a positive definite* G^c-*invariant kernel on* $\Xi \times \Xi$ *for which the canonical representation* $(U, \mathscr{H}_\Psi)$ *of* G^c *is irreducible, then there exist* $c > 0$ *and* $\lambda \in i\mathbb{R} \cup (0, \frac{n-1}{2}]$ *such that* $\Psi = c\Phi_\lambda$ *and* $(U, \mathscr{H}_\Psi)$ *is equivalent to* $(U^\lambda, \mathscr{E}^\lambda)$.

Proof The function Ψ_{e_0} is K-invariant. Hence Theorem 7.3.13 implies that there exists a $\lambda \in i\mathbb{R} \cup (0, \frac{n-1}{2}]$ such that $(L, \mathscr{H}_\Psi) \simeq (U^\lambda, \mathscr{H}_\lambda)$. We can assume that $\Psi(e_0, e_0) = 1$. Then $\Psi(e_0, g.e_0) = \langle \Psi_{e_0}, \Psi_{g.e_0} \rangle = \langle \Psi_{e_0}, L_g \Psi_{e_0} \rangle = \langle 1, U^\lambda_g 1 \rangle = \varphi_\lambda(g)$.   □

The Boundary of the Crown and the Spherical Function φ_λ

In this subsection we describe the boundary of the crown as a disjoint union of two orbits. Both are homogeneous spaces that we have already met, the de Sitter space dS^n and the forward pointing light like vectors $\mathbb{L}^n_+$ which we have already introduced in Sects. 7.3.3 and 7.3.1, respectively. For details we refer to [NÓ18].

A simple calculation shows that the boundary of Ξ in $\mathbb{S}^n_\mathbb{C}$ is given by

$$\partial \Xi = \left\{ z = u + iv \in V + iV = \mathbb{C}^{n+1} : \begin{array}{l} [u, u]_V = 0, u_0 \geq 0, \\ [v, v]_V = -1, [u, v]_V = 0 \end{array} \right\} \tag{7.31}$$

([NÓ18, Lemma 3.10]). If $u = 0$, then (7.31) leads to a realization of *de Sitter space*

$$dS^n := i\{v \in V : [v, v]_V = -1\} = iV \cap \mathbb{S}^n_\mathbb{C} = G^c.ie_n \cong G^c/H \subseteq \partial \Xi \cap iV$$

where $H \simeq O_{1,n-1}(\mathbb{R})$ is the stabilizer of $e_n \in dS^n$.

Let $\xi_0 := e_0 + ie_n$ and write $G^c_{\xi_0} = MN$, where M and N are similar to the groups introduced in Sect. 7.3.1 except one has to replace v by iv and interchange the second and last columns and second and last row as we now consider e_n as a base point.

Lemma 7.4.15 *Suppose that* $n \geq 2$ *and let* $\mathcal{O} := G^c.(e_0 + ie_n + e_{n-1})$. *The boundary of the crown is the union two* G^c-*orbits*

$$\partial \Xi = dS^n \,\dot\cup\, \mathcal{O}.$$

In particular, dS^n *is the unique open* G^c-*orbit in the boundary. The projection of* $\mathcal{O}$ *onto* V *is* $\mathbb{L}^n_+$ *and the projection onto* iV *is* dS^n.

The orbit $G^c.ie_n \cong dS^n$ is called the *Shilov boundary* of Ξ in $\mathbb{S}^n_\mathbb{C}$. The tangent space of dS^n at e_n is the n-dimensional Minkowski space

$$T_{e_n}(\mathrm{dS}^n) = i\mathbb{R} \oplus \mathbb{R}^{n-1} \simeq \mathbb{R}^{1,n-1}.$$

By (7.23), we have

$$\mathrm{Exp}_{e_n}(z) = S(z^2)z + C(z^2)e_n \quad \text{for} \quad z \in T_{e_n}(\mathrm{dS}^n)_{\mathbb{C}} = \mathbb{C} \oplus \mathbb{C}^{n-1}. \tag{7.32}$$

We now describe how one can obtain the crown by moving inward from the de Sitter space dS^n.

Theorem 7.4.16 *For $g \in G$ let $V_{+,g.e_n}^{\pi} = \{v \in g.V_+ : [v, v]_V < \pi^2\} \subset T_{g.e_n}(\mathrm{dS}^n)$. Then*

$$\Xi = G^c.\,\mathrm{Exp}_{e_n}(V_{+,e_n}^{\pi}) = \bigcup_{p \in \mathrm{dS}^n} \mathrm{Exp}_p(V_{+,p}^{\pi}).$$

Proof In view of the G^c-invariance of Ξ and the equivariance of the exponential map of $\mathbb{S}_{\mathbb{C}}^n$, it suffices to verify the first equality. From (7.32) we obtain for $v \in \mathbb{R}_+ \oplus i\mathbb{R}^{n-1} \subset iT_{e_n}(\mathrm{dS}^n)$ and $\mathbb{R}_+ = (0, \infty)$:

$$\mathrm{Exp}_{e_n}(v) = S([v, v]_V)v + C([v, v]_V)e_n \tag{7.33}$$

and this is contained in $T_{V_+} \cap \mathbb{S}_{\mathbb{C}}^n = \Xi$ if $[v, v]_V \in (0, \pi^2)$. Therefore $\mathrm{Exp}_{e_n}(\Omega_{e_n}) \subseteq \Xi$. If, conversely, $z \in \Xi = G^c.\mathbb{S}_+^n$, there exists a $t \in (0, \pi)$ such that z is G^c-conjugate to $x = (\sin t, 0, \ldots, 0, \cos t)$. But then $te_0 \in V_{+,e_n}^{\pi}$, and (7.33) yields $x = \mathrm{Exp}_{e_n}(te_0)$. This proves the claim. $\qquad\square$

In this section we give a different description of the spherical function φ_λ and the kernel $\Phi_\lambda(x, y)$ (cf. Sect. 7.4.3) using the space $\mathbb{L}_+^n$. For that we have to assume that $m \geq \frac{n-1}{2}$ (which corresponds to the principal series), which we do from now on.

Recall the map $\xi : \mathbb{S}^{n-1} \to \mathbb{L}_+^n = G^c.\xi_0, x \mapsto (1, x)$ and the action of G^c on $\mathbb{S}^{n-1} \cong \mathbb{L}_+^n / \mathbb{R}_+^{\times}$, given by

$$\xi(g.u) = J(g, u)^{-1} g(\xi(u)).$$

Lemma 7.4.17 *Let $z \in \Xi$ and $\xi \in \mathbb{L}_+^n$. Then $\mathrm{Re}\,[z, \xi]_V > 0$.*

Proof Write $z = u + iv \in \Xi \subseteq T_{V_+}$ (Proposition 7.4.8). Then $u \in V_+$ implies that

$$\mathrm{Re}\,[z, \xi_0]_V = [u, \xi_0]_V = u_0 - u_n > 0 \quad \text{for} \quad \xi_0 = e_0 + ie_n \in V.$$

But then $\mathrm{Re}\,[z, g.\xi_0]_V = \mathrm{Re}\,[g^{-1}.z, \xi_0]_V > 0$ for all $g \in G^c$. $\qquad\square$

For $\lambda \in \mathbb{C}$ we define the analytic kernel

$$K_\lambda : \Xi \times \mathbb{L}_+^n \to \mathbb{C}, \qquad K_\lambda(z, \xi) := K_{\lambda,\xi}(z) := [z, \xi]_V^{\lambda - \frac{n-1}{2}}. \tag{7.34}$$

This kernel is continuous for $\mathrm{Re}\,\lambda > (n-1)/2$. Note the similarity with the distribution vector p_λ from (7.19).

Theorem 7.4.18 *For $\lambda \in i[0, \infty) \cup (0, (n-1)/2)$, the assignment*

$$\mathscr{P}_\lambda : L^2(\mathbb{S}^{n-1}) \to \mathscr{O}(\Xi), \quad (\mathscr{P}_\lambda \varphi)(z) := \int_{\mathbb{S}^{n-1}} K_\lambda(z, \xi(u)) \varphi(u) \, d\mu_{\mathbb{S}^{n-1}}(u)$$

defines a G^c-intertwining operator $(U^\lambda, L^2(\mathbb{S}^{n-1})) \to (L, \mathscr{H}_{\Phi_\lambda})$ with $\mathscr{P}_\lambda 1 = \varphi_\lambda$.

Notes

Most of Sects. 7.1–7.3 is from [NÓ14], with slightly different notation. Under some additional assumptions, Theorem 7.1.7 can be found in [vD09, Theorem 8.2.1], and in [NÓ14, Sect. 2] for the special case $H = \{e\}$. In Sect. 7.3 we added some material on the principal series representations and the H-invariant distributions vectors. The notation in Sect. 7.3 has also been adapted to the standard notation from [JÓl98, JÓl00, Ól00, FÓØ18] as well as the notation for the last section in this chapter. The material about the sphere is from [NÓ18]; for its relation to construction of QFTs on de Sitter space, we refer to [BJM16].

The crown is the maximal G^c-invariant domain for the holomorphic extension of all spherical functions on the Riemannian symmetric space G^c/K. It is shown in [KS05], that in our case the crown is a Riemannian symmetric space $\mathrm{SO}_{2,n}(\mathbb{R})_0/(\mathrm{SO}_2(\mathbb{R}) \times \mathrm{SO}_n(\mathbb{R}))$; see also [NÓ18] for a direct argument. For more information about the crown see [AG90, KO08, KS04].

Chapter 8
Generalized Free Fields

We now turn to representations of the Poincaré group corresponding to scalar generalized free fields and their euclidean realizations by representations of the euclidean motion group. We start in Sect. 8.1 with a brief discussion of Lorentz invariant measures on the forward light cone $\overline{V_+}$ and turn in Sect. 8.2 to the corresponding unitary representations. Applying the dilation construction to the time translation semigroup leads immediately to a euclidean Hilbert space $\mathscr{E}$ on which we have a unitary representation of the euclidean motion group. In Sect. 8.3 we characterize those representations which extend to the conformal group $O_{2,d}(\mathbb{R})$ of Minkowski space. Then the euclidean realization is a unitary representation of the Lorentz group $O_{1,d+1}(\mathbb{R})$, acting as the conformal group on euclidean $\mathbb{R}^d$.

8.1 Lorentz Invariant Measures on the Light Cone and Their Relatives

Before we turn to unitary representations of the Poincaré group, it is instructive to have a closer look at Lorentz invariant measures μ on the forward light cone $\overline{V_+}$ and their projections to $\mathbb{R}^{d-1}$. We shall also see that these measures are directly related to rotation invariant measures ν on euclidean space $\mathbb{R}^d$, and this establishes the key link between unitary representations of the Poincaré group $P(d)$ and the euclidean group $E(d)$.

Definition 8.1.1 For $m \geq 0$ or $d > 1$, we define a Borel measure μ_m on

$$H_m := \{p \in \mathbb{R}^d : [p, p] = p_0^2 - \mathbf{p}^2 = m^2, \ p_0 > 0\}$$
$$\subseteq \overline{V_+} = \{p = (p_0, \mathbf{p}) \in \mathbb{R}^d : p_0 \geq 0, \ [p, p] = p_0^2 - \mathbf{p}^2 \geq 0\}$$

© The Author(s) 2018
K.-H. Neeb and G. Ólafsson, *Reflection Positivity*, SpringerBriefs in Mathematical
Physics, https://doi.org/10.1007/978-3-319-94755-6_8

by

$$\int_{\mathbb{R}^d} f(p)\,d\mu_m(p) = \int_{\mathbb{R}^{d-1}} f\left(\sqrt{m^2 + \mathbf{p}^2}, \mathbf{p}\right) \frac{d\mathbf{p}}{\sqrt{m^2 + \mathbf{p}^2}}$$

(cf. [RS75, Chap. IX], [vD09, Lemma 9.1.2/3]). These measures are invariant under the Lorentz group $O_{1,d-1}(\mathbb{R})^\uparrow$ and every Lorentz invariant measure μ on $\overline{V_+}$ is of the form

$$\mu = c\delta_0 + \int_0^\infty \mu_m\,d\rho(m), \tag{8.1}$$

where $c \geq 0$ and ρ is a Borel measure on $[0, \infty)$ (with $\rho(\{0\}) = 0$ for $d = 1$) whose restriction to $\mathbb{R}_+$ is a Radon measure (see [NÓ15a, Theorem B.1]).[1]

Remark 8.1.2 (a) For $d = 1$, we have $H_m = \{m\}$ for $m > 0$ and $H_0 = \emptyset$. Therefore μ_0 does not make sense. For $m > 0$, we have $\mu_m = \frac{1}{m}\delta_m$, where δ_m is the Dirac measure in m.

(b) For $d = 2$, the measure μ_0 is singular in 0, but every $\varphi \in \mathscr{S}(\mathbb{R}^2)$ vanishing in 0 is integrable (cf. [GJ81, p. 103]). In particular, this measure does not define a distribution, it defines a functional on the smaller space of test functions $\mathscr{S}_*(\mathbb{R}^d) := \{\varphi \in \mathscr{S}(\mathbb{R}^d): \varphi(0) = 0\}$.

(c) By [NÓ15a, Theorem B.1], the measure μ in (8.1) is tempered if and only if the measure ρ is tempered and, in addition,

$$\int_0^1 \frac{1}{m}\,d\rho(m) < \infty \quad \text{for} \quad d = 1, \qquad \int_0^1 \ln(m^{-1})\,d\rho(m) < \infty \quad \text{for} \quad d = 2. \tag{8.2}$$

Example 8.1.3 (Generalized free fields)

(a) For the scalar generalized free field of spin zero on $\mathbb{R}^d$, the corresponding one-particle Hilbert space is $\mathscr{H} := L^2(\mathbb{R}^d, \mu)$, where μ is a Lorentz invariant measure on $\overline{V_+}$ (see (8.1)). Here the *time translation semigroup* C_t acts by the contractions

$$(C_t f)(p) = e^{-tp_0} f(p).$$

The dilation construction from Example 4.3.8 leads to the space $\mathscr{E} := L^2(\mathbb{R}^{d+1}, \zeta)$ with

$$d\zeta(\lambda, p) = \frac{1}{\pi}\frac{p_0}{p_0^2 + \lambda^2}\,d\lambda\,d\mu(p) \quad \text{and} \quad (U_t f)(\lambda, p) = e^{it\lambda} f(\lambda, p). \tag{8.3}$$

For $\mathrm{pr}_2(\lambda, p_0, \mathbf{p}) = (\lambda, \mathbf{p})$, the projected measure $\nu := (\mathrm{pr}_2)_*\zeta$ on $\mathbb{R}^d$ is given, in the special case $\mu = \mu_m$, by the measure ν_m from Example 2.4.7:

---

[1] In Quantum Field Theory this is known as the Lehmann Spectral Formula for two-point functions; see [GJ81, Theorem 6.2.4].

$$dv_m(p_0, \mathbf{p}) := \frac{1}{\pi} \frac{\sqrt{m^2 + \mathbf{p}^2}}{m^2 + \mathbf{p}^2 + p_0^2} \, dp_0 \frac{1}{\sqrt{m^2 + \mathbf{p}^2}} \, d\mathbf{p} = \frac{1}{\pi} \frac{d\mathbf{p}}{m^2 + p^2}, \qquad (8.4)$$

so that

$$dv(p_0, \mathbf{p}) = \frac{1}{\pi} \int_0^\infty \frac{1}{m^2 + p^2} \, dp \, d\rho(m) = \left(\frac{1}{\pi} \int_0^\infty \frac{d\rho(m)}{m^2 + p^2} \right) dp = \Theta(p) \, dp,$$

$$(8.5)$$

for $\Theta(p) := \frac{1}{\pi} \int_0^\infty \frac{d\rho(m)}{m^2 + p^2}$.

(b) Since elements of $L^2(\mathbb{R}^d, v)$ correspond to functions in $L^2(\mathbb{R}^{d+1}, \zeta)$ not depending on the second argument p_0, we obtain an isometric embedding

$$\text{pr}_2^* \colon L^2(\mathbb{R}^d, v) \to \mathcal{E} = L^2(\mathbb{R}^{d+1}, \zeta). \qquad (8.6)$$

(c) The free scalar field of mass m and spin $s = 0$ on $\mathbb{R}^d$ (with $m > 0$ or $d > 1$) corresponds to the measure $\mu = \mu_m$ (cf. [GJ81, p. 103]). In this case pr_2^* is surjective, so that we can identify $L^2(\mathbb{R}^d, v)$ with $\mathcal{E}$. The measure v_m is finite if and only if $d = 1$ and $m > 0$. It is tempered if and only if $d > 2$ or $m > 0$.

Definition 8.1.4 We call a positive Borel measure ρ on $[0, \infty)$ *tame* if $\int_0^\infty \frac{d\rho(m)}{1+m^2} < \infty$. Note that this implies in particular that ρ is tempered.

Remark 8.1.5 In view of [NÓ15a, Lemma 7.1], the measure ρ is tame if and only if $\Theta(p) < \infty$ for every $p \in \mathbb{R}^d$ and this in turn is equivalent to $L^2(\mathbb{R}^d, v) \neq \{0\}$.

If this is the case, then the measure v on $\mathbb{R}^d$ is tempered if and only if $d > 2$ or the conditions (8.2) characterizing the temperedness of μ for $d = 1, 2$ are satisfied ([NÓ15b, Proposition 7.3]). As tameness of ρ implies that ρ is tempered, μ is tempered if v has this property.

8.2 From the Poincaré Group to the Euclidean Group

We have already seen in Example 8.1.3 that Lorentz invariant measures on the forward light cone lead by the dilation construction to rotation invariant measures on euclidean space. We now take a closer look of the implications of this correspondence for unitary representations of the Poincaré group $P(d)$ and the euclidean group $E(d)$. In QFT, this corresponds to the one-particle representations of scalar generalized free fields.

Example 8.2.1 (One particle representation of generalized free fields) Let μ be a Lorentz invariant Radon measure as in (8.1) on the forward light cone $\overline{V_+} \subseteq \mathbb{R}^d$ with $c = \mu(\{0\}) = 0$. Then we have a natural unitary representation of the Poincaré group $G^c := P(d)^\uparrow = \mathbb{R}^d \rtimes O_{1,d-1}(\mathbb{R})^\uparrow$ on

$$\mathcal{H} := L^2(\overline{V_+}, \mu) \quad \text{by} \quad (U^\mu(x, g)f)(p) := e^{ixp} f(g^{-1}p).$$

Analytic continuation of the time-translation group leads to the contraction semi-group

$$(C_t f)(p) = (U^\mu(ite_0, \mathbf{1})f)(p) = e^{-tp_0} f(p),$$

and the dilation construction from Example 4.3.8, applied to this contraction semigroup, leads to the Hilbert space

$$\mathcal{E} = L^2(\mathbb{R} \times \overline{V_+}, \zeta) = L^2(\mathbb{R}^{d+1}, \zeta) \quad \text{with} \quad d\zeta(\lambda, p) = \frac{1}{\pi} \frac{p_0}{p_0^2 + \lambda^2} \, d\lambda \, d\mu(p)$$

(cf. Example 8.1.3).

We consider the unitary representation U of the euclidean translation group of $\mathbb{R}^d$ on $\mathcal{E}$, given by

$$\big(U(x_0, \mathbf{x})f\big)(\lambda, p_0, \mathbf{p}) = e^{-i(x_0\lambda + \mathbf{x}\mathbf{p})} f(\lambda, p_0, \mathbf{p}). \tag{8.7}$$

The constant function 1 on $\mathbb{R}^{d+1}$ is a distribution vector for U if and only if the projected measure $\nu = (\mathrm{pr}_2)_* \zeta$ under $\mathrm{pr}_2(x, p_0, \mathbf{p}) = (x, \mathbf{p})$, it tempered (cf. Remark 8.1.5 for criteria), and then the corresponding distribution is $D = \widehat{\nu}$ (Lemma 7.1.9).

It is remarkable that the measure ν on $\mathbb{R}^d$ is rotation invariant, so that dilation with respect to the contraction semigroup $(C_t)_{t \geq 0}$ leads directly from the representation U^μ of the Poincaré group on $L^2(\mathbb{R}^d, \mu)$ to a representation U^ν of the euclidean motion group $E(d)$ on $L^2(\mathbb{R}^d, \nu)$ by

$$(U^\nu(x, g)f)(p) := e^{-ixp} f(g^{-1}p).$$

For $\mu = \mu_m$, the representation U^{μ_m} of the Poincaré group is irreducible because the measure μ_m lives on a single $\mathrm{O}_{1,d-1}(\mathbb{R})^\uparrow$-orbit in $\mathbb{R}^d$ (it is $\mathrm{O}_{1,d-1}(\mathbb{R})^\uparrow$-ergodic). As the measure ν_m is a proper superposition of the invariant measures on spheres of any radius, the corresponding representation U^{ν_m} of $E(d)$ is reducible and a direct integral of representations corresponding to the invariant measures on the spheres of radius r. Since all measures $(\nu_m)_{m>0}$ are equivalent to Lebesgue measure, all representations $(U^{\nu_m})_{m>0}$ are actually equivalent.

Proposition 8.2.2 *If ρ is a tame measure on $[0, \infty)$ for which the measure ν is tempered, then the rotation invariant distribution $\widehat{\nu} \in C^{-\infty}(\mathbb{R}^d)$ is reflection positive for $(\mathbb{R}^d, \mathbb{R}^d_+, \theta)$ and $\theta(x_0, \mathbf{x}) = (-x_0, \mathbf{x})$, i.e.,*

$$\int_{\mathbb{R}^d} \overline{\widetilde{\psi}} \cdot \theta \widehat{\psi} \, d\nu \geq 0 \quad for \quad f \in C_c^\infty(\mathbb{R}^d_+).$$

Proof (see also [GJ81, Proposition 6.2.5]) Writing $\nu = \int_0^\infty \nu_m \, d\rho(m)$ with $d\nu_m(p) = \frac{1}{\pi} \frac{dp}{m^2 + p^2}$, the assertion follows from the reflection positivity of the distributions $\widehat{\nu_m}$ verified in Example 2.4.7.   □

We now assume that the measure ν is tempered (cf. Remark 8.1.5). Then the corresponding distribution $D = \widehat{\nu}$ is reflection positive by Proposition 8.2.2. Let

$$\mathscr{F} := \mathrm{pr}_2^*(L^2(\mathbb{R}^d, \nu)) \subseteq \mathscr{E} = L^2(\mathbb{R}^{d+1}, \zeta)$$

be the image under the isometry pr_2^* from (8.6). It coincides with $[\![U_{C_c^\infty(\mathbb{R}^d)}1]\!]$ (see 8.7), and reflection positivity of $\widehat{\nu}$ implies that the subspace $\mathscr{F}_+ := [\![U_{C_c^\infty(\mathbb{R}_+^d)}1]\!]$ is θ-positive.

To see how this fits the subspace $\mathscr{E}_0$ and $\mathscr{E}_+$ of $\mathscr{E}$, we first note that

$$\mathscr{F}_0 := \mathscr{F} \cap \mathscr{E}_0$$

consists of those function in $\mathscr{E}_0$ that are also independent of p_0. This is the L^2-space of the projected measure $\widetilde{\nu} := (\mathrm{pr}_1)_*\nu$ on $\mathbb{R}^{d-1}$ for $\mathrm{pr}_1(p_0, \mathbf{p}) = \mathbf{p}$. Since $\nu = (\mathrm{pr}_2)_*\zeta$, we also have $\widetilde{\nu} := (\mathrm{pr}_1)_*\mu$. According to [NÓ15a, Theorem B.1], $\mathscr{F}_0 \cong L^2(\mathbb{R}^{d-1}, \widetilde{\nu})$ is non-zero if and only if the measure $\widetilde{\nu}$ is tempered, which is equivalent to the additional condition

$$\int_1^\infty \frac{d\rho(m)}{m} < \infty \tag{8.8}$$

on the growth of ρ at infinity. Assume that $\widetilde{\nu}$ is tempered. Then 1 is a distribution vector for the representation $U|_{\mathbb{R}^{d-1}}$ on $\mathscr{E}$, and the corresponding cyclic subspace coincides with $\mathscr{F}_0 \subseteq \mathscr{E}_0$ (Lemma 7.1.9). This in turn implies that $\mathscr{F}_+ \subseteq \mathscr{E}_+$. Further, $\widehat{\mathscr{E}} \cong L^2(\mathbb{R}^d, \mu)$ contains the subspace $\widehat{\mathscr{F}}_0 \cong \mathscr{F}_0 \cong L^2(\mathbb{R}^{d-1}, \widetilde{\nu})$ of functions not depending on p_0, and the canonical map $\mathscr{F}_0 \to \widehat{\mathscr{E}_0}$ is unitary. Accordingly, the "time zero-subspace" $\widehat{\mathscr{F}}_0$ is the same on the euclidean and the Minkowski side.

Since $\mathscr{F}_0$ is U-cyclic in $\mathscr{F}$, the subspace $\widehat{\mathscr{F}}_0$ is $\widehat{U}$-cyclic in $\widehat{\mathscr{E}}$, showing that $\widehat{\mathscr{F}} = \widehat{\mathscr{E}}$. Therefore the representation U^ν of the euclidean group $E(d)$ on $\mathscr{F}$ provides a euclidean realization of the representation $(U^\mu, L^2(\mathbb{R}^d, \mu))$ of $P(d)^\uparrow$. To see how $\mathscr{F}$ is generated from $\mathscr{F}_0$, we now determine the corresponding positive definite operator-valued function

$$\varphi: \mathbb{R} \to B(\mathscr{F}_0), \quad \varphi(t) = P_0 U(t, 0)P_0^*,$$

where $P_0: \mathscr{E} \to \mathscr{F}_0$ is the orthogonal projection. This function is determined by the relation

$$\langle \xi, \varphi(t)\eta \rangle = \langle \xi, U_t\eta \rangle \quad \text{for} \quad \xi, \eta \in \mathscr{F}_0.$$

We have

$$\langle \xi, \varphi(t)\eta \rangle = \int_{\mathbb{R}^d} e^{-itp_0}\overline{\xi(\mathbf{p})}\eta(\mathbf{p})\, d\nu(\mathbf{p}) = \int_{\mathbb{R}^d} e^{-itp_0}\overline{\xi(\mathbf{p})}\eta(\mathbf{p})\, \Theta(p_0, \mathbf{p})\, dp_0\, d\mathbf{p}$$

$$= \int_{\mathbb{R}^{d-1}} \overline{\xi(\mathbf{p})}\eta(\mathbf{p}) \int_{\mathbb{R}} e^{-itp_0}\Theta(p_0, \mathbf{p})\, dp_0\, d\mathbf{p} = \int_{\mathbb{R}^{d-1}} \overline{\xi(\mathbf{p})}\eta(\mathbf{p})\Theta_t(\mathbf{p})\, d\mathbf{p},$$

where (8.5) yields

$$
\begin{aligned}
\Theta_t(\mathbf{p}) &:= \int_{\mathbb{R}} e^{-itp_0} \Theta(p_0, \mathbf{p})\, dp_0 = \frac{1}{\pi} \int_{\mathbb{R}} \int_0^\infty e^{-itp_0} \frac{m}{m^2 + \mathbf{p}^2 + p_0^2}\, d\rho(m)\, dp_0 \\
&= \int_0^\infty \left(\frac{1}{\pi} \int_{\mathbb{R}} e^{-itp_0} \frac{m}{(m^2 + \mathbf{p}^2) + p_0^2}\, dp_0 \right) d\rho(m) \\
&= \int_0^\infty \frac{m}{\sqrt{m^2 + \mathbf{p}^2}} e^{-|t|\sqrt{m^2 + \mathbf{p}^2}}\, d\rho(m).
\end{aligned}
$$

Here we have used Example 2.4.3 in the calculation. Now $d\widetilde{\nu}(\mathbf{p}) = \Theta_0(\mathbf{p})d\mathbf{p}$ implies that the operator $\varphi(t)$ on $\mathscr{F}_0$ is given by multiplication with the function Θ_t/Θ_0.

For the subspace $\mathscr{F}_0 \subseteq \widehat{\mathscr{E}}$ and $f, g \in \mathscr{F}_0$, the relation

$$
\langle \widehat{f}, \widehat{U}_t \widehat{g} \rangle = \langle f, \theta U_t g \rangle = \langle f, U_t g \rangle = \langle f, \varphi(t) g \rangle = \langle \widehat{f}, \varphi(t) \widehat{g} \rangle
$$

implies that $\varphi|_{\mathbb{R}_+}$ is the positive definite function on $\mathbb{R}_+$ corresponding to the cyclic subspace $\mathscr{F}_0 \subseteq \widehat{\mathscr{E}}$.

Example 8.2.3 For the special case where $\rho = \delta_m$ with $m > 0$ or $d > 2$, we have

$$
\Theta_t(\mathbf{x}) = \frac{m}{\sqrt{m^2 + \mathbf{x}^2}} e^{-|t|\sqrt{m^2 + \mathbf{x}^2}}, \quad \text{and} \quad \frac{\Theta_t(\mathbf{x})}{\Theta_0(\mathbf{x})} = e^{-|t|\sqrt{m^2 + \mathbf{x}^2}}
$$

is multiplicative for $t \geq 0$. This corresponds to the fact that $q(\mathscr{E}_0) = \widehat{\mathscr{E}}$ (the Markov case; Proposition 3.4.9), which in turn is due to the fact that the inclusion $L^2(\mathbb{R}^{d-1}, \widetilde{\nu}_m) \hookrightarrow L^2(\mathbb{R}^d, \mu_m)$ is surjective.

This has the interesting consequence that, if we consider elements of $\widehat{\mathscr{E}}$ as functions

$$
f : \mathbb{R}_+ \times \mathbb{R}^{d-1} \to \mathbb{C}
$$

as in the preceding example, we have

$$
f(t, \mathbf{p}) = (\widehat{U}_t f)(0, \mathbf{p}) = e^{-t\sqrt{m^2 + \mathbf{p}^2}} f(0, \mathbf{p}). \tag{8.9}
$$

This in turn leads by analytic continuation to

$$
f(it, \mathbf{p}) = (U_t^c f)(0, \mathbf{p}) = e^{it\sqrt{m^2 + \mathbf{p}^2}} f(0, \mathbf{p}). \tag{8.10}
$$

These formulas provide rather conceptual direct arguments for formulas like [GJ81, Proposition 6.2.5].

Remark 8.2.4 A unitary representation $(U, \mathscr{H})$ of the Poincaré group is said to be of *positive energy* if the spectrum of the time translation group is non-negative. In view of the covariance with respect to the Lorentz group $O_{1,d-1}(\mathbb{R})^\uparrow$, this is equivalent to the spectral measure of $U|_{\mathbb{R}^d}$ to be supported in the closed forward light cone $\overline{V_+}$ because this is the set of all orbits of $O_{1,d-1}(\mathbb{R})^{\uparrow}$ on which the function p_0 is non-negative.

If such a representation is multiplicity free on $\mathbb{R}^d$, then $\mathscr{H} \cong L^2(\overline{V_+}, \mu)$ for a measure μ on $\overline{V_+}$ which is quasi-invariant under $O_{1,d-1}(\mathbb{R})^{\uparrow}$. Since the action of $O_{1,d-1}(\mathbb{R})^{\uparrow}$ on $\overline{V_+}$ has a measurable cross section and every orbit carries an invariant measure, the measure μ can be chosen $O_{1,d-1}(\mathbb{R})^{\uparrow}$-invariant. The representation U is irreducible if and only if the measure μ is ergodic, i.e., $\mu = \mu_m$ for some $m \geq 0$ (with $m > 0$ for $d = 1$) or $\mu = \delta_0$ (the Dirac measure in 0).

For all the multiplicity free representations $(U^{\mu}, L^2(\overline{V_+}, \mu))$, Example 8.2.1 provides a euclidean realization in the dilation space $\mathscr{E} = L^2(\mathbb{R}_+ \times \overline{V_+}, \zeta)$, as far as the representation of the subgroup $\mathbb{R}^d \rtimes O_{d-1}(\mathbb{R})$ is concerned. The subspace $\mathscr{E}_0 \subseteq \mathscr{E}$ is invariant under the subgroup $G^{\tau} \cong \mathbb{R}^{d-1} \rtimes O_{d-1}(\mathbb{R})$, which also implies the invariance of $\mathscr{E}_+$ under this group.

A euclidean realization for the full group is obtained in Example 8.1.3 for irreducible representations, i.e., $\mu = \mu_m$. In the general case we assume that ν is tempered. Then the following theorem is the bridge between the reflection positive representation U^{ν} of $E(d)$ on $\mathscr{F} \cong L^2(\mathbb{R}^d, \nu)$ and the representation $U^c \cong U^{\mu}$ of the Poincaré group on $\widehat{\mathscr{F}} \cong L^2(\mathbb{R}^d, \mu)$.

Theorem 8.2.5 *If ν is tempered, then $1 \in \mathscr{E}^{-\infty}$ is a reflection positive distribution vector for the representation U of $\mathbb{R}^d$. Accordingly, we obtain a reflection positive representation of $\mathbb{R}^d$ on the subspace $\mathscr{F} \subseteq \mathscr{E}$ generated by $U^{-\infty}(C_c^{\infty}(\mathbb{R}^d))1$. The corresponding reflection positive distribution $\widehat{\nu}$ on $\mathbb{R}^d$ is rotation invariant, so that $\mathscr{F}$ carries a reflection positive representation of $E(d)$ for which $\mathscr{F}_0$ and $\mathscr{F}_+$ are invariant under $H := E(d)^{\tau} \cong \mathbb{R}^{d-1} \rtimes O_{d-1}(\mathbb{R})$.*

Moreover, $\widehat{\mathscr{F}} \cong L^2(\overline{V_+}, \mu)$, $q: \mathscr{F}_+ \to \widehat{\mathscr{F}}$ is H-equivariant and we have the relation $\widehat{U}(t, 0) = U^{\mu}(it, 0, 1)$ for the positive energy representation U^{μ} of the Poincaré group $P(d)^{\uparrow}$ on $\widehat{\mathscr{F}}$.

Proof We have already seen that $1 \in \mathscr{E}^{-\infty}$ is equivalent to ν being tempered (Lemma 7.1.9). To determine the corresponding space $\widehat{\mathscr{F}}$, we have to take a closer look at the corresponding reflection positive distribution $D = \widehat{\nu}$ for $(\mathbb{R}^d, \mathbb{R}^d_+, \theta)$ (Proposition 8.2.2). In view of [NÓ14, Proposition 2.12], this follows if we can show that $D|_{\mathbb{R}^d_+}$ coincides with the Fourier–Laplace transform

$$\mathscr{F}\mathscr{L}(\mu)(x) := \int_{\mathbb{R}^d_+} e^{-x_0 p_0} e^{i\mathbf{x}\mathbf{p}} \, d\mu(p).$$

First we observe that the temperedness of μ implies that $\mathscr{F}\mathscr{L}(\mu)(x)$ exists pointwise and defines an analytic function on $\mathbb{R}^d_+$. Here the main point is that, on $\overline{V_+}$ we have $p^2 = p_0^2 + \mathbf{p}^2 \leq 2p_0^2$ (cf. [NÓ14, Example 4.12]). We have

$$\mathscr{F}\mathscr{L}(\mu)(x) = \int_{V_+} e^{-x_0 p_0} e^{i x \mathbf{p}} \, d\mu(p) = \int_0^\infty \int_{\mathbb{R}^{d-1}} e^{-x_0 p_0} e^{i x \mathbf{p}} \, d\mu(p_0, \mathbf{p})$$

$$= \int_0^\infty \int_{\mathbb{R}^{d-1}} \left(\frac{1}{\pi} \int_{\mathbb{R}} e^{i t x_0} \frac{p_0}{p_0^2 + t^2} \, dt \right) e^{i x \mathbf{p}} \, d\mu(p_0, \mathbf{p})$$

$$= \int_{\mathbb{R} \times \mathbb{R}^d} e^{i(t x_0 + x \mathbf{p})} \, d\zeta(t, p_0, \mathbf{p}) = \int_{\mathbb{R}^d} e^{i(t x_0 + x \mathbf{p})} \, d\nu(t, \mathbf{p}) = \widehat{\nu}(x).$$

If μ is infinite, then the triple integral only exists as an iterated integral in the correct order, not in the sense that the integrand is Lebesgue integrable. One can deal with this problem by integrating against a test function on $\mathbb{R}^d_+$, and then the above calculation shows that $\mathscr{F}\mathscr{L}(\mu)$ coincides with $\widehat{\nu}$ on $\mathbb{R}^d_+$ as a distribution.   □

8.3 The Conformally Invariant Case

In this section we study the special case where the measure μ on $\overline{V_+}$ is semi-invariant under homotheties. This provides a bridge to the complementary series representations of $O_{1,d+1}(\mathbb{R})^\uparrow$ discussed in Sect. 7.3.2 because then the representation of $E(d)$ on $L^2(\mathbb{R}^d, \nu)$ extends to the conformal group $O_{1,d+1}(\mathbb{R})$ of $\mathbb{R}^d$.

Lemma 8.3.1 ([NÓ15b, Lemma 5.17]) *An $O_{1,d-1}(\mathbb{R})^\uparrow$-invariant measure $\mu = \int_0^\infty \mu_m \, d\rho(m)$ on $\overline{V_+}$ is semi-invariant under homotheties if and only if*

$$d\rho(m) = m^{s-1} \, dm \quad for\ some \quad s \in \mathbb{R}.$$

If this is the case, then ρ is tempered if and only if $s > 0$, and μ is tempered if $d > 1$ or $s > 1$. For $d > 1$, the measure $\widetilde{\nu}$ on $\mathbb{R}^{d-1}$ is tempered if and only if $s > 1$. For $d = 1$, the measure μ is never finite.

From now on we write $d\rho_s(m) = m^{s-1} \, dm$ on $[0, \infty)$. As the measure μ is semi-invariant under homotheties, we can expect the corresponding representation of the Poincaré group to extend to the conformal group $SO_{2,d}(\mathbb{R})$ of Minkowski space.

Lemma 8.3.2 ([NÓ15b, Lemma 5.18], Proposition 7.3.10) *The measure $\nu = \Theta_s \cdot dp$ corresponding to ρ_s is tempered if and only if $0 < s < 2$ for $d > 1$ and if $0 < s < 1$ for $d = 1$. In this case Θ_s is a multiple of $\|p\|^{s-2}$ and the Fourier transform $\widehat{\nu}$ is a positive multiple of $\|x\|^{-d+2-s}$.*

The preceding lemma implies in particular that the distribution $\|x\|^{-a}$ on $\mathbb{R}^d$ is reflection positive for $d - 2 < a < d$, which has been obtained in [NÓ14, Proposition 6.1], [FL10, Lemma 2.1] and [FL11, Lemma 3.1] by other means. This connection is made more precise in the following theorem:

Theorem 8.3.3 *Let* $0 < s < 2$, *resp.,* $1 < s < 2$ *for* $d = 1$. *Then*

(i) The canonical representation of the conformal motion group

$$CE(d) := \mathbb{R}^d \rtimes (O_d(\mathbb{R}) \times \mathbb{R}_+^\times)$$

on $\mathscr{E} := L^2(\mathbb{R}^d, \nu) \cong \mathscr{H}_D$ *for* $D(x) = \|x\|^{-d+2-s}$ *extends to a complementary series representation of the orthochronous euclidean conformal group* $O_{1,d+1}(\mathbb{R})_+$.

(ii) The corresponding representation of the conformal Poincaré group

$$CP(d)^\uparrow := \mathbb{R}^d \rtimes (L^\uparrow \times \mathbb{R}_+^\times)$$

is irreducible and extends to a representation of a covering of the relativistic conformal group $SO_{2,d}(\mathbb{R})_0$.

Proof (i) From Lemma 8.3.2 we know that $\mathscr{E} := L^2(\mathbb{R}^d, \nu)$ can be identified with the Hilbert space $\mathscr{H}_D$ obtained by completion of $C_c^\infty(\mathbb{R}^d)$ with respect to the scalar product

$$\langle \varphi, \psi \rangle_s := \int_{\mathbb{R}^d} \int_{\mathbb{R}^d} \frac{\overline{\varphi(x)}\psi(y)}{\|x - y\|^{d-2+s}} \, dx \, dy$$

(cf. Definition 2.4.5). Now Theorem 7.3.8 implies that the representation of $E(d)$ on this space extends to an irreducible complementary series representation of the conformal group $O_{1,d+1}(\mathbb{R})_+$

(ii) The irreducibility of the representation U^c follows from the transitivity of the action of $\mathbb{R}_+^\times O_{1,d-1}(\mathbb{R})^\uparrow$ on the open light cone V_+. To see that this representation extends to $SO_{2,d}(\mathbb{R})_0$, we can use the fact that the representation U of the conformal group of $\mathbb{R}^d$ is reflection positive with respect to the open subsemigroup of strict compressions of the open half space $\mathbb{R}_+^d$ in the conformal compactification $\mathbb{S}^d$. As explained in [JÓl00, Sects. 6, 10], see also [HN93], [JÓl98], the reflection positivity and the Lüscher–Mack Theorem now provide an irreducible representation of the simply connected c-dual group G^c on $\widehat{\mathscr{E}}$.   □

Chapter 9
Reflection Positivity and Stochastic Processes

In this chapter we describe some recent generalizations of classical results by Klein and Landau [Kl78, KL75] concerning the interplay between reflection positivity and stochastic processes. Here the main step is the passage from the symmetric semigroup $(\mathbb{R}, \mathbb{R}_+, -\mathrm{id}_\mathbb{R})$ to more general triples (G, S, τ). This leads to the concept of a (G, S, τ)-measure space generalizing Klein's Osterwalder–Schrader path spaces for $(\mathbb{R}, \mathbb{R}_+, -\mathrm{id}_\mathbb{R})$. A key result is the correspondence between (G, S, τ)-measure spaces and the corresponding positive semigroup structures on the Hilbert space $\widehat{\mathcal{E}}$.

The exposition in this chapter is minimal in the sense that the main results are explained and full definitions are given. For more details we refer to [JN15].

9.1 Reflection Positive Group Actions on Measure Spaces

We start with the basic concepts related to (G, S, τ)-measure spaces which provide a measure theoretic perspective on reflection positive representations of symmetric semigroups (G, S, τ).

Definition 9.1.1 Let (G, τ) be a symmetric group. A (G, τ)-*measure space* is a quadruple $((Q, \Sigma, \mu), \Sigma_0, U, \theta)$ consisting of the following ingredients:

(GP1) a measure space (Q, Σ, μ),
(GP2) a sub-σ-algebra Σ_0 of Σ,
(GP3) a measure preserving action $U : G_\tau \to \mathrm{Aut}(\mathscr{A})$ on the von Neumann algebra $\mathscr{A} := L^\infty(Q, \Sigma, \mu)$, for which the corresponding unitary representation on $L^2(Q, \mu)$ is continuous, and
(GP4) $\theta = U_\tau$ satisfies $\theta E_0 \theta = E_0$, where $E_0 : L^\infty(Q, \Sigma, \mu) \to L^\infty(Q, \Sigma_0, \mu)$ is the conditional expectation.
(GP5) Σ is generated by the sub-σ-algebras $\Sigma_g := U_g \Sigma_0, g \in G$.

If μ is a probability measure, we speak of a (G, τ)-*probability space*. If $S = S^\sharp \subseteq G$ is a symmetric subsemigroup, then we write $\Sigma_\pm$ for the sub-σ-algebra generated by $(\Sigma_s)_{s \in S^{\pm 1}}$, and $E_\pm$ for the corresponding conditional expectations.

© The Author(s) 2018
K.-H. Neeb and G. Ólafsson, *Reflection Positivity*, SpringerBriefs in Mathematical Physics, https://doi.org/10.1007/978-3-319-94755-6_9

Definition 9.1.2 (a) A (G, τ)-measure space is called *reflection positive with respect to the symmetric subsemigroup S* if

$$\langle \theta f, f \rangle \geq 0 \quad \text{for} \quad f \in \mathscr{E}_+ := L^2(Q, \Sigma_+, \mu).$$

This is equivalent to $E_+ \theta E_+ \geq 0$ as an operator on $L^2(Q, \Sigma, \mu)$ and obviously implies $\theta E_0 = E_0$. If this condition is satisfied and, in addition, Σ_0 is invariant under the unit group $H(S) := S \cap S^{-1}$, then we call it a (G, S, τ)-*measure space*.[1]

(b) A *Markov (G, S, τ)-measure space* is a (G, S, τ)-measure space with the *Markov property* $E_+ E_- = E_+ E_0 E_-$ (cf. Definition 2.3.1).

Proposition 3.4.9 immediately provides a reflection positive representation on the corresponding L^2-space:

Proposition 9.1.3 *For a (G, S, τ)-measure space $((Q, \Sigma, \mu), \Sigma_0, U, \theta)$, we put $\mathscr{E} := L^2(Q, \Sigma, \mu)$, $\mathscr{E}_0 := L^2(Q, \Sigma_0, \mu)$ and $\mathscr{E}_\pm := L^2(Q, \Sigma_\pm, \mu)$. Then the natural action of G on $\mathscr{E}$ defines a reflection positive representation of (G, S, τ).*

The Markov property is equivalent to the natural map $\mathscr{E}_0 \to \widehat{\mathscr{E}}$ being unitary and this implies that the positive definite function $\varphi \colon S \to B(\mathscr{E}_0)$, $\varphi(s) = E_0 U_s E_0$ is multiplicative and the unitary representation U of G on $(\mathscr{E}, \mathscr{E}_+, \theta)$ is a euclidean realization of the $*$-representation $(\varphi, \mathscr{E}_0)$ of $(S, \sharp)$.

Example 9.1.4 Typical examples arise in QFT as follows. Let $\mathscr{E}$ be a real Hilbert space and $X = \mathscr{E}^*$ be its algebraic dual space, i.e., the space of all linear functionals $\mathscr{E} \to \mathbb{R}$, continuous or not. On this set we consider the smallest σ-algebra for which all evaluation functionals $\varphi(\xi)(\alpha) := \alpha(\xi)$ are measurable. Then there exists a Gaussian measure μ on X such that any tuple $(\varphi(\xi_1), \ldots, \varphi(\xi_n))$ is jointly Gaussian with covariance $(\langle \xi_i, \xi_j \rangle)_{1 \leq i, j \leq n}$ [JN15, Example 4.3], [Sim05, Theorem 2.3.4]. The orthogonal group $O(\mathscr{E})$ acts in a measure preserving way on X by $U\alpha := \alpha \circ U^{-1}$.

If we start with a reflection positive unitary representation $(U, \mathscr{E}, \mathscr{E}_+, \theta)$ of (G, S, τ), for which $\mathscr{E}_0$ is U-cyclic and $\mathscr{E}_+$ is generated by $U_S \mathscr{E}_0$, then all this structure is reflected in (X, Σ, μ). In particular, we obtain a measure preserving action of G_τ on X. We write $\Sigma_0 \subseteq \Sigma$ for the smallest σ-subalgebra for which all functions $(\varphi(\xi))_{\xi \in \mathscr{E}_0}$ are measurable. Then Σ_+ is generated by the translates $(U_s \Sigma_0)_{s \in S}$ and (GP1-5) are satisfied.

The following concept aims at an axiomatic characterization of the corresponding semigroup representations on the spaces $\widehat{\mathscr{E}}$. It generalizes the corresponding classical concepts for the case $(G, S, \tau) = (\mathbb{R}, \mathbb{R}_+, -\mathrm{id}_\mathbb{R})$ ([Kl78] for (a) and [KL75] for (b)).

Definition 9.1.5 (a) A *positive semigroup structure* for a symmetric semigroup (G, S, τ) is a quadruple $(\mathscr{H}, P, \mathscr{A}, \Omega)$ consisting of

(PS1) a Hilbert space $\mathscr{H}$,

---

[1] Note that $E_+ \theta E_+ \geq 0$ is equivalent to the kernel $K^\theta(A, B) := \mu(A \cap \theta(B))$ on Σ_+ being positive definite, i.e., the kernel $K(A, B) := \mu(A \cap B)$ on Σ is reflection positive with respect to $(\Sigma, \Sigma_+, \theta)$ (Definition 2.4.1).

(PS2) a strongly continuous $*$-representation $(P_s)_{s \in S}$ of $(S, \sharp)$ by contractions on $\mathcal{H}$,

(PS3) a commutative von Neumann algebra $\mathscr{A}$ on $\mathcal{H}$ normalized by the operators $(P_h)_{h \in S \cap S^{-1}}$, and

(PS4) a unit vector $\Omega \in \mathcal{H}$, such that

(i) $P_s \Omega = \Omega$ for every $s \in S$.

(ii) Ω is cyclic for the (not necessarily selfadjoint) subalgebra $\mathscr{B} \subseteq B(\mathcal{H})$ generated by $\mathscr{A}$ and $\{P_s : s \in S\}$.

(iii) For positive elements $A_1, \ldots, A_n \in \mathscr{A}$ and $s_1, \ldots, s_{n-1} \in S$, we have

$$\langle \Omega, A_1 P_{s_1} A_2 \cdots P_{s_{n-1}} A_n \Omega \rangle \geq 0.$$

(b) A *standard positive semigroup structure* for a symmetric semigroup (G, S, τ) consists of a σ-finite measure space $(M, \mathfrak{S}, \nu)$ and

(SPS1) a representation $(P_s)_{s \in S}$ of S on $L^\infty(M, \nu)$ by positivity preserving operators, i.e., $P_s f \geq 0$ for $f \geq 0$.

(SPS2) $P_s 1 = 1$ for $s \in S$ (the Markov condition).

(SPS3) P is involutive with respect to ν, i.e., $\int_M P_s(f) h \, d\nu = \int_M f P_{s^\sharp}(h) \, d\nu$ for $s \in S$, $f, h \geq 0$.

(SPS4) P is strongly continuous in measure, i.e., for each $f \in L^1(M, \nu) \cap L^\infty(M, \nu)$ and every $\delta > 0$, $s_0 \in S$, we have $\lim_{s \to s_0} \nu(\{|P_s f - P_{s_0} f| \geq \delta\}) = 0$.

The main difference between these two concepts is that (b) concerns the situation where $\mathcal{H}$ is an L^2-space, but it also leaves some additional freedom because the measure ν is not required to be finite so that the constant function 1 need not be L^2.

The following proposition shows that the requirement that Ω is cyclic for $\mathscr{A}$ describes those positive semigroup structures which are standard.

Proposition 9.1.6 *Let $(M, \mathfrak{S}, \nu)$ be a probability space and $(P_s)_{s \in S}$ be a positivity preserving continuous $*$-representation of $(S, \sharp)$ by contractions on $L^2(M, \nu)$, i.e.,*

$$P_s 1 = 1 \quad and \quad P_s f \geq 0 \quad for \quad f \geq 0, s \in S.$$

Then $(L^2(M), Q, L^\infty(M), 1)$ is a standard positive semigroup structure for which 1 is a cyclic vector for $L^\infty(M)$.

Conversely, let $(\mathcal{H}, P, \mathscr{A}, \Omega)$ be a positive semigroup structure for which Ω is a cyclic vector for $\mathscr{A}$. Then there exists a probability space M and a positivity preserving semigroup $(Q_s)_{s \in S}$ on $L^2(M)$ such that $(\mathcal{H}, P, \mathscr{A}, \Omega) \cong (L^2(M), Q, L^\infty(M), 1)$ as positive semigroup structures.

The following theorem characterizes the positive semigroup structures arising in the Markov context as those for which Ω is a cyclic vector for $\mathscr{A}$, which is considerably stronger than condition (PS4)(b).

Theorem 9.1.7 *Let* $((Q, \Sigma, \mu), \Sigma_0, U, \theta)$ *be a* (G, S, τ)*-probability space and let* $(\widehat{\mathscr{E}}, \widehat{U}, \mathscr{A}, \Omega)$ *be its associated positive semigroup structure. Then* $((Q, \Sigma, \mu), \Sigma_0, U, \theta)$ *is Markov if and only if* Ω *is* $\mathscr{A}$*-cyclic in* $\widehat{\mathscr{E}}$.

Proof The Markov property is equivalent to $q(\mathscr{E}_0) = \widehat{\mathscr{E}}$ (Proposition 9.1.3). Since $\mathscr{A} \cdot 1$ is dense in $\mathscr{E}_0$, this is equivalent to $\Omega = q(1)$ being $\mathscr{A}$-cyclic in $\widehat{\mathscr{E}}$.   □

Example 9.1.8 (*The real oscillator semigroup*) We consider the Hilbert space $\mathscr{H} = L^2(\mathbb{R}^d)$, with respect to Lebesgue measure.

(a) On $\mathscr{H}$ we have a unitary representation by the group $\mathrm{GL}_d(\mathbb{R})$ by

$$(T_h f)(x) := |\det(h)|^{-d/2} f(h^{-1}x) \quad \text{for} \quad h \in \mathrm{GL}_d(\mathbb{R}), x \in \mathbb{R}^d,$$

and we also have two representations of the additive abelian semigroup $\mathrm{Sym}_d(\mathbb{R})_+$ (the convex cone of positive semidefinite matrices):

(1) Each $A \in \mathrm{Sym}_d(\mathbb{R})_+$ defines a multiplication operator $(M_A f)(x) := e^{-\langle Ax,x \rangle} f(x)$ which is positivity preserving on $L^\infty(\mathbb{R}^n)$ but does not preserve 1; it preserves the Dirac measure δ_0 in the origin.
(2) Each $A \in \mathrm{Sym}_d(\mathbb{R})_+$ specifies a uniquely determined (possibly degenerate) Gaussian measure μ_A on $\mathbb{R}^d$ whose Fourier transform is given by $\widehat{\mu}_A(x) = e^{-\langle Ax,x \rangle/2}$. Then the convolution operator $C_A f := f * \mu_A$ is positivity preserving and leaves Lebesgue measure on $\mathbb{R}^d$ invariant. For $A = \mathbf{1}$, we thus obtain the heat semigroup as $(\mu_{t1})_{t \geq 0}$.

Any composition of these 3 types of operators T_h, M_A and C_A is positivity preserving on $L^\infty(\mathbb{R}^d)$, and they generate a $*$-representation of the Olshanski semigroup $S := H \exp(C)$ in the symmetric Lie group $G := \mathrm{Sp}_{2d}(\mathbb{R})$, where $H = \mathrm{GL}_d(\mathbb{R})$, $C = \mathrm{Sym}_d(\mathbb{R})_+ \times \mathrm{Sym}_d(\mathbb{R})_+ \subseteq \mathrm{Sym}_d(\mathbb{R}) \oplus \mathrm{Sym}_d(\mathbb{R}) = \mathfrak{q}$, and

$$\tau \begin{pmatrix} A & B \\ C & -A^\top \end{pmatrix} = \begin{pmatrix} A & -B \\ -C & -A^\top \end{pmatrix} \quad \text{for} \quad \begin{pmatrix} A & B \\ C & -A^\top \end{pmatrix} \in \mathfrak{sp}_{2d}(\mathbb{R}) \text{ with } B^\top = B, C^\top = C$$

(cf. Examples 3.2.6). The real Olshanski semigroup S is the fixed point set of an antiholomorphic involutive automorphism of the so-called oscillator semigroup $S_{\mathbb{C}} = G^c \exp(iW)$ which is a complex Olshanskii semigroup [How88, Hi89]. The elements in the interior of S act on $L^2(\mathbb{R}^d)$ by kernel operators with positive Gaussian kernels and the elements of $S_{\mathbb{C}}$ correspond to complex-valued Gaussian kernels. The semigroup S contains many interesting symmetric one-parameter semigroups such as the Mehler semigroup $e^{-t H_{\mathrm{osc}}}$ generated by the oscillator Hamiltonian

$$H_{\mathrm{osc}} = -\sum_{j=1}^n \partial_j^2 + \frac{1}{4} \sum_{j=1}^n x_j^2 - \frac{d}{2} \mathbf{1} \tag{9.1}$$

which fixes the Gaussian $e^{-\|x\|^2/4}$.

(b) The subsemigroup $S := \mathrm{Sym}_d(\mathbb{R})_+ \rtimes \mathrm{GL}_d(\mathbb{R}) \subseteq \mathrm{Sp}_{2d}(\mathbb{R})$ also is a symmetric subsemigroup of (G, τ) with $G = \mathrm{Sym}_d(\mathbb{R}) \rtimes \mathrm{GL}_d(\mathbb{R})$ and $\tau(A, g) = (-A, g)$. Here the commutative von Neumann algebra $\mathscr{A} = L^\infty(\mathbb{R}^d)$ is invariant under conjugation with the operators T_h, so that $(A, h) \mapsto C_A T_h$ defines a $*$-representation of $(S, \sharp)$ that leads to a standard positive semigroup structure on $L^2(\mathbb{R}^d)$.

9.2 Stochastic Processes Indexed by Lie Groups

We now introduce stochastic processes where the more common index set $\mathbb{R}$ is replaced by a Lie group G. The forward direction is then given by a subsemigroup S of G. So called stationary stochastic processes correspond naturally to measure preserving G-actions on spaces G^Q of all maps $Q \to G$.

Definition 9.2.1 Let (Q, Σ, μ) be a probability space. A *stochastic process* indexed by a group G is a family $(X_g)_{g \in G}$ of measurable functions $X_g \colon Q \to (B, \mathfrak{B})$, where $(B, \mathfrak{B})$ is a measurable space, called the *state space* of the process. It is called *full* if, up to sets of measure zero, Σ is the smallest σ-algebra for which all functions $(X_g)_{g \in G}$ are measurable.

For such a process, we obtain a measurable map

$$\Phi \colon Q \to B^G, \quad \Phi(q) = (X_g(q))_{g \in G}$$

with respect to the product σ-algebra $\mathfrak{B}^G$. Then $\nu := \Phi_*\mu$ is a measure on B^G, called the *distribution of the process* $(X_g)_{g \in G}$. This measure is uniquely determined by the measures $\nu_{\mathbf{g}}$ on G^n, obtained for any finite tuple $\mathbf{g} := (g_1, \ldots, g_n) \in G^n$ as the image of μ under the map $X_{\mathbf{g}} = (X_{g_1}, \ldots, X_{g_n}) \colon Q \to B^n$ (cf. [Hid80, Sect. 1.3]). If $\mathbf{g} = (g)$ for some $g \in G$, then we write ν_g for $\nu_{\mathbf{g}}$.

The process $(X_g)_{g \in G}$ is called *stationary* if the corresponding distribution on B^G is invariant under the translations

$$(U_g \nu)_h := \nu_{g^{-1}h} \quad \text{for} \quad g, h \in G.$$

If $\tau \in \mathrm{Aut}(G)$ is an automorphism, then we call the process τ-*invariant* if its distribution is invariant under $(\tau \nu)_h := \nu_{\tau^{-1}(h)}$ for $h \in G$.

The connection with (G, S, τ)-measure spaces is now easily described:

Example 9.2.2 Let (G, τ) be a symmetric Lie group and $(X_g)_{g \in G}$ be a stationary, τ-invariant, full stochastic process on (Q, Σ_Q, μ_Q). Then its distribution $(B^G, \mathfrak{B}^G, \nu)$ satisfies the conditions (GP1, 2, 4, 5) of a (G, τ)-probability space with respect to the canonical actions of G and τ on B^G, where Σ_0 is the σ-algebra generated by $(X_h)_{h \in G^\tau}$, i.e., the smallest subalgebra for which these functions are measurable. In this context (GP3) is equivalent to the continuity of the unitary representation of G on $L^2(B^G, \mathfrak{B}^G, \nu)$.

9.3 Associated Positive Semigroup Structures and Reconstruction

The main result of this section is the Reconstruction Theorem. It asserts that, if $G = S \cup S^{-1}$, positive semigroup structures come from (G, S, τ)-measure spaces. For a subsemigroup $S \subseteq G$, we consider the left invariant *partial order* $\prec_S$ on G defined by $g \prec_S h$ if $g^{-1}h \in S$, i.e., $h \in gS$.

Lemma 9.3.1 *Let* $((Q, \Sigma, \mu), \Sigma_0, U, \theta)$ *be a* (G, S, τ)-*measure space. Consider the von Neumann algebra* $\mathscr{A} := L^\infty(Q, \Sigma_0, \mu)$ *on*

$$\mathscr{E} := L^2(Q, \Sigma, \mu) \supseteq \mathscr{E}_+ := L^2(Q, \Sigma_+, \mu) \supseteq \mathscr{E}_0 := L^2(Q, \Sigma_0, \mu),$$

and the canonical map $q: \mathscr{E}_+ \to \widehat{\mathscr{E}}$. *Then the following assertions hold:*

(a) *For* $f \in \mathscr{A}$, *let* M_f *denote the corresponding multiplication operator on* $\mathscr{E}$. *Then there exists a bounded operator* $\widehat{M}_f \in B(\widehat{\mathscr{E}})$ *with* $q \circ M_f|_{\mathscr{E}_+} = \widehat{M}_f \circ q$ *and* $\|\widehat{M}_f\| = \|f\|_\infty$.

(b) $U(f) := \widehat{M}_f$ *is a faithful weakly continuous representation of the commutative von Neumann algebra* $\mathscr{A}$ *on* $\widehat{\mathscr{E}}$.

(c) *In the Markov case we identify* $\widehat{\mathscr{E}}$ *with* $\mathscr{E}_0$ *and* q *with* E_0 (*Proposition* 9.1.3). *For* $g_1 \prec_S g_2 \prec_S \cdots \prec_S g_n$ *in* G, *non-negative functions* $f_1, \ldots, f_n \in \mathscr{A}$ *and* $f_{g_j} := U_{g_j} f_j$, *we have*

$$\int_Q f_{g_1} \cdots f_{g_n} \, d\mu = \int_Q \widehat{M}_{f_1} \widehat{U}_{g_1^{-1}g_2} \cdots \widehat{M}_{f_{n-1}} \widehat{U}_{g_{n-1}^{-1}g_n} \widehat{M}_{f_n} 1 \, d\mu.$$

If, in addition, μ *is finite, then* $\Omega := \mu(Q)^{-1/2} q(1)$ *satisfies:*

(d) *For* $g_1 \prec_S g_2 \prec_S \cdots \prec_S g_n$ *in* G, $f_1, \ldots, f_n \in \mathscr{A}$ *and* $f_{g_j} := U_{g_j} f_j$, *we have*

$$\int_Q f_{g_1} \cdots f_{g_n} \, d\mu = \Big\langle \widehat{M}_{f_1} \widehat{U}_{g_1^{-1}g_2} \cdots \widehat{M}_{f_{n-1}} \widehat{U}_{g_{n-1}^{-1}g_n} \widehat{M}_{f_n} \Omega, \Omega \Big\rangle.$$

(e) Ω *is a separating vector for* $\mathscr{A}$ *and* $\widehat{U}_s \Omega = \Omega$ *for every* $s \in S$.

(f) Ω *is cyclic for the algebra* $\mathscr{B}$ *generated by* $\mathscr{A}$ *and* $(\widehat{U}_s)_{s \in S}$.

Definition 9.3.2 The preceding lemma shows that, if $((Q, \Sigma, \mu), \Sigma_0, U, \theta)$ is a finite (G, S, τ)-measure space, then $(\widehat{\mathscr{E}}, \widehat{U}, \mathscr{A}, q(1))$ is a positive semigroup structure for $\mathscr{A} = \{\widehat{M}_f : f \in L^\infty(Q, \Sigma_0, \mu)\}$. We call it the *associated positive semigroup structure*.

We now turn to our version of Klein's Reconstruction Theorem. Note that every discrete group is in particular a 0-dimensional Lie group, so that the following theorem applies in particular to discrete groups.

Theorem 9.3.3 (Reconstruction Theorem) *Let (G, S, τ) be a symmetric semigroup satisfying $G = S \cup S^{-1}$. Then the following assertions hold:*

(a) *Every positive semigroup structure for (G, S, τ) is associated to some (G, S, τ)-probability space $((Q, \Sigma, \mu), \Sigma_0, U, \theta)$.*

(b) *Every standard positive semigroup structure for (G, S, τ) is associated to some (G, S, τ)-measure space $((Q, \Sigma, \mu), \Sigma_0, U, \theta)$ which is unique up to G-equivariant isomorphism of measure spaces.*

Remark 9.3.4 Without going into details of the proof, it is instructive to take a closer look at the construction of the measure space (Q, Σ, μ) in the proof of Theorem 9.3.3 in [JN15]. Here G acts unitarily on the Hilbert space $\mathcal{H} \cong L^2(M, \mathfrak{S}, \nu)$. For simplicity, we assume that $(M, \mathfrak{S}, \nu)$ (cf. Proposition 9.1.6) is a polish space, i.e., M carries a topology for which it is completely metrizable and separable and $\mathfrak{S}$ is the σ-algebra of Borel sets. Then [Ba96, Corollary 35.4] implies the existence of a Borel measure μ on the measurable space $(M^G, \mathfrak{S}^G)$ with the projections onto finite products satisfying

$$\int_{M^G} f_1(\omega(g_1)) \cdots f_n(\omega(g_n)) \, d\mu(\omega) = \int_Q M_{f_1} P_{g_1^{-1}g_2} \cdots M_{f_{n-1}} P_{g_{n-1}^{-1}g_n} M_{f_n} 1 \, d\nu$$

for $0 \leq f_1, \ldots, f_n \in L^\infty(M, \mathfrak{S}, \nu)$ and $g_1 \prec_S \cdots \prec_S g_n$. We thus obtain a realization of our (G, S, τ)-measure space on $(M^G, \mathfrak{S}^G, \mu)$, where the measure preserving G-action on M^G is given by $(g.\omega)(h) := \omega(g^{-1}h)$.

Definition 9.3.5 ([Ba78]) (a) Let (Q, Σ) and (Q', Σ') be measurable spaces. Then a function $K : Q \times \Sigma' \to [0, \infty]$ is called a *kernel* if

(K1) for every $A' \in \Sigma'$, the function $K^{A'}(\omega) := K(\omega, A')$ is Σ-measurable, and

(K2) for every $\omega \in Q$, the function $K_\omega(A') := K(\omega, A')$ is a (positive) measure.

A kernel is called a *Markov kernel* if the measures K_ω are probability measures.

(b) A kernel $K : Q \times \Sigma' \to [0, \infty]$ associates to a measure μ on (Q, Σ) the measure

$$(\mu K)(A') := \int \mu(d\omega) K(\omega, A') = \int_Q K(\omega, A') \, d\mu(\omega).$$

(c) If $(Q_j, \Sigma_j)_{j=1,2,3}$ are measurable spaces, then composition of kernels K_1 on $Q_1 \times \Sigma_2$ and K_2 on $Q_2 \times \Sigma_3$ is defined by $(K_1 K_2)(\omega_1, A_3) = \int K_1(\omega_1, d\omega_2) K_2(\omega_2, A_3)$. If S is a semigroup, then a family $(P_s)_{s \in S}$ of Markov kernels on the measurable space (Q, Σ) is called a *semigroup of (Markov) kernels* if $P_s P_t = P_{st}$ for $st \in S$.

Remark 9.3.6 (a) Let $(P_t)_{t \geq 0}$ be a Markov semigroup on the polish space (Q, Σ). Then we obtain for $0 \leq t_1 < \ldots < t_n$ and $\mathbf{t} = (t_1, \ldots, t_n)$ a Markov kernel $P_{\mathbf{t}}$ on $Q \times \Sigma^n$ by

$$P_{\mathbf{t}}(x_0, B) = \int_{Q^n} \chi_B(x_1, \dots, x_n) P_{t_1}(x_0, dx_1) P_{t_2-t_1}(x_1, dx_2) \cdots P_{t_n-t_{n-1}}(x_{n-1}, dx_n)$$

[Ba78, Satz 64.2]. Fixing x_0, we thus obtain a a projective family of measures, which leads to a probability measure P_{x_0} on the σ-algebra $\Sigma^{\mathbb{R}_+}$ on $Q^{\mathbb{R}_+}$ whose restrictions to cylinder sets are given by the $P_{\mathbf{t}}(x_0, \cdot)$. We thus obtain a Markov kernel $P(x, \cdot) := P_x(\cdot)$ on $Q \times \Sigma^{\mathbb{R}_+}$. For any $t \geq 0$, we then have

$$P_t(x, B) = \int_{Q^{\mathbb{R}_+}} \chi_B(\omega(t)) P(x, d\omega) = P(x, \{\omega(t) \in B\}),$$

which leads to

$$(P_t f)(x) = \int_Q P_t(x, dy) f(y) = \int_{Q^{\mathbb{R}_+}} P(x, d\omega) f(\omega(t)). \tag{9.2}$$

This is an abstract version of the Feynman–Kac–Nelson formula that expresses the value of $P_t f$ in $x \in Q$ as an integral over all paths $[0, t] \to Q$ starting in x with respect to the probability measure P_x.

(b) For any measure ν on Q, we thus obtain a measure $P^\nu := \nu P$ on $(Q^{\mathbb{R}_+}, \Sigma^{\mathbb{R}_+})$. If ν is a probability measure, then P^ν likewise is, and we obtain a stochastic process $(X_t)_{t \geq 0}$ with state space (Q, Σ) and initial distribution ν [Ba78, Satz 62.3]. According to [Ba78, Satz 65.3], the so obtained stochastic process has the Markov property.

For $t > 0$, we have the relation

$$\int_{Q_+^{\mathbb{R}}} f(\omega(t)) \, dP^\nu(\omega) = \int_Q \nu(dx) P_t(x, dy) f(y) \quad \text{for} \quad t \in \mathbb{R},$$

and, for $t < s$,

$$\int_Q \int_Q f_1(x) \nu(dx) P_{s-t}(x, dy) f_2(y) = \int_{Q^{\mathbb{R}_+}} f_1(\omega(t)) f_2(\omega(s)) \, dP^\nu(\omega).$$

(c) In the special case where $Q = G$ is a topological group and $P_t f = f * \mu_t$ for probability measures μ_t on G, we have

$$(P_t f)(x) = \int_G f(xy) \, d\mu_t(y) = \int_G P_t(x, dy) f(y) \quad \text{for} \quad P_t(x, A) = \mu_t(x^{-1}A).$$

Let $P(G) := G^{\mathbb{R}}$ be the *path group* of all maps $\mathbb{R} \to G$ and let $P_*(G)$ be the subgroup of pinned paths $P_*(G) = \{\omega \in P(G) : \omega(0) = e\}$. We have the relations

$$\int_{P(G)} f_1(\omega(0)) f_2(\omega(t)) \, dP^\nu(\omega) = \int_G \int_G f_1(g_1) f_2(g_1 g_2) \, d\nu(g_1) d\mu_t(g_2).$$

for $t > 0$, and

$$\int_{P_*(G)} f(\omega(t)) \, dP^\nu(\omega) = \int_G \int_G f(xg) \, d\nu(x) d\mu_t(g) = \int_G f(g) \, d(\nu * \mu_t)(g).$$

This leads for $f \in L^2(G, \nu)$ and $t \geq 0$ to

$$(P_t f)(x) = (f * \mu_t)(x) = \int_G f(xg) \, d\mu_t(g) = \int_{G^{\mathbb{R}_+}} f(x\omega(t)) \, dP(\omega).$$

This is a group version of the Feynman–Kac–Nelson formula (9.2).

We now assume that G is a second countable locally compact group and that $(\mu_t)_{t \geq 0}$ is a convolution semigroup of probability measures on G which is *strongly continuous* in the sense that $\lim_{t \to 0} \mu_t = \delta_e = \mu_0$ weakly on the space $C_b(G)$ of bounded continuous functions on G. We further assume that ν is a measure on G satisfying $\nu * \mu_t = \nu$ for every $t > 0$, and, in addition, that the operators $P_t f :=$ $f * \mu_t$ on $L^2(G, \nu)$ are symmetric. If ν is a right Haar measure, then the symmetry of the operators P_t is equivalent to $\mu_t^* = \mu_t$. Then we obtain for $t_1 \leq \cdots \leq t_n$ and $\mathbf{t} := (t_1, \dots, t_n)$ on G^n a consistent family of measures

$$P_{\mathbf{t}}^\mu := (\psi_n)_* (\nu \otimes \mu_{t_2 - t_1} \otimes \cdots \otimes \mu_{t_n - t_{n-1}}),$$

where $\psi_n(g_1, \dots, g_n) = (g_1, g_1 g_2, \cdots, g_1 \cdots g_n)$. This in turn leads to a unique measure P^ν on the two-sided path space $G^{\mathbb{R}}$ with $(\mathrm{ev}_{\mathbf{t}})_* P^\nu = P_{\mathbf{t}}^\nu$ for $t_1 < \cdots < t_n$.

From the Klein–Landau Reconstruction Theorem we obtain the following specialization.

Theorem 9.3.7 *Suppose that G is a second countable locally compact group. Let P^ν be the measure on $G^{\mathbb{R}}$ corresponding to the convolution semigroup $(\mu_t)_{t \geq 0}$ of symmetric probability measures on G and the measure ν on G for which the operators $P_t f = f * \mu_t$ define a positive semigroup structure on $L^2(G, \nu)$. Then the translation action $(U_t \omega)(s) := \omega(s - t)$ on $G^{\mathbb{R}}$ is measure preserving and P^ν is invariant under $(\theta \omega)(t) := \omega(-t)$. We thus obtain a reflection positive one-parameter group of Markov type on*

$$\mathscr{E} := L^2(G^{\mathbb{R}}, \mathfrak{B}^{\mathbb{R}}, \mu) \quad \text{with respect to} \quad \mathscr{E}_+ := L^2(G^{\mathbb{R}}, \mathfrak{B}^{\mathbb{R}_+}, \mu),$$

for which $\mathscr{E}_0 := \mathrm{ev}_0^*(L^2(G, \nu)) \cong L^2(G, \nu)$ and $\widehat{\mathscr{E}} \cong L^2(G, \nu)$ with $q(F) = E_0 F$ for $F \in \mathscr{E}_+$. We further have

$$E_0 U_t E_0 = P_t \quad \text{for} \quad P_t f = f * \mu_t,$$

so that the U-cyclic subrepresentation of $\mathscr{E}$ generated by $\mathscr{E}_0$ is a unitary dilation of the one-parameter semigroup $(P_t)_{t \geq 0}$ of hermitian contractions on $L^2(G, \nu)$.

Example 9.3.8 (a) For $G = \mathbb{R}^d$, the heat semigroup is given on $L^2(\mathbb{R}^d)$ by

$$e^{t\Delta} f = f * \gamma_t \quad \text{where} \quad d\gamma_t(x) = \frac{1}{(2\pi t)^{d/2}} e^{-\frac{\|x\|^2}{2t}} \, dx.$$

We call the corresponding measure on $G^{\mathbb{R}}$ the *Lebesgue–Wiener measure* (cf. Theorem 9.3.7).

(b) If G is any finite dimensional Lie group and $X_1, \ldots, X_n$ is a basis of the Lie algebra, then we obtain a left invariant *Laplacian* by $\Delta := \sum_{j=1}^n L_{X_j}^2$, where L_{X_j} denotes the left invariant vector field with $L_{X_j}(e) = X_j$. Then there also exists a semigroup $(\mu_t)_{t \geq 0}$ of probability measures on G such that $e^{t\Delta} f = f * \mu_t$ for $t \geq 0$ [Nel69, Sect. 8]. Accordingly, we obtain a *Haar–Wiener measure* on the path space $G^{\mathbb{R}}$.

Notes

The material in this section is condensed from [JN15] to which we refer for more details and background. This paper draws heavily from the work of Klein and Landau [KL75, Kl78]. In particular, the Markov (G, S, τ)-measure spaces generalize the path spaces studied by Klein and Landau in [KL75]. For $(G, S, \tau) = (\mathbb{R}, \mathbb{R}_+, - \mathrm{id}_{\mathbb{R}})$, the work of Klein and Landau was largely motivated by Nelson's work on the Feynman–Kac Formula in [Ne00].

In A. Klein's papers [Kl77, Kl78] concerning $(G, S, \tau) = (\mathbb{R}, \mathbb{R}_+, - \mathrm{id})$, the reflection positivity condition from Definition 9.1.2 is called Osterwalder–Schrader positivity. Theorem 9.1.7 is adapted from [Kl78, Theorem 3.1].

Stochastic processes index by Lie groups also appear in [AHH86].

Appendix
Background Material

In this appendix we collect precise statements of some basic facts on positive definite kernels and positive definite functions on groups and semigroups.

A.1 Positive Definite Kernels

Let X be a set. Classically, reproducing kernels arise from Hilbert spaces $\mathscr{H}$ which are subspaces of the space $\mathbb{C}^X$ of complex-valued functions on X, for which the evaluations $f \mapsto f(x)$ are continuous, hence representable by elements $K_x \in \mathscr{H}$ by

$$f(x) = \langle K_x, f \rangle \quad \text{for} \quad f \in \mathscr{H}, x \in X.$$

Then

$$\cdot K : X \times X \to \mathbb{C}, \quad K(x, y) := K_y(x) = \langle K_x, K_y \rangle$$

is called the *reproducing kernel* of $\mathscr{H}$. As the kernel K determines $\mathscr{H}$ uniquely, we write $\mathscr{H}_K \subseteq \mathbb{C}^X$ for the Hilbert space determined by K and $\mathscr{H}_K^0 \subseteq \mathscr{H}_K$ for the subspace spanned by the functions $(K_x)_{x \in X}$. A kernel function $K : X \times X \to \mathbb{C}$ is the reproducing kernel of some Hilbert space if and only if it is *positive definite* in the sense that, for any finite collection $x_1, \ldots, x_n \in X$, the matrix $(K(x_j, x_k))_{1 \le j,k \le n}$ is positive semidefinite (cf. [Ar50], [Nel64, Chap. 1]). There is a natural generalization to Hilbert spaces $\mathscr{H}$ of functions with values in a Hilbert space $\mathscr{V}$, i.e., $\mathscr{H} \subseteq \mathscr{V}^X$. Then $K_x(f) = f(x)$ is a linear operator $K_x : \mathscr{H} \to \mathscr{V}$ and we obtain a kernel $K(x, y) := K_x K_y^* \in B(\mathscr{V})$ with values in the bounded operators on $\mathscr{V}$. However, there are also situations where one would like to deal with kernels whose values are unbounded operators, so that one has to generalize this context further. The notion of a positive definite kernel with values in the space $\mathrm{Bil}(V)$ of bilinear complex-valued forms on a real linear space V provides a natural context to deal with all relevant cases.

K.-H. Neeb and G. Ólafsson, *Reflection Positivity*, SpringerBriefs in Mathematical Physics, https://doi.org/10.1007/978-3-319-94755-6

Definition A.1.1 Let X be a set and V be a real vector space. We write $\mathrm{Bil}(V) = \mathrm{Bil}(V, \mathbb{C})$ for the space of complex-valued bilinear forms on V. We call a map $K: X \times X \to \mathrm{Bil}(V)$ *a positive definite kernel* if the associated scalar-valued kernel

$$K^{\flat}: (X \times V) \times (X \times V) \to \mathbb{C}, \quad K^{\flat}((x, v), (y, w)) := K(x, y)(v, w)$$

is positive definite.

The corresponding reproducing kernel Hilbert space $\mathcal{H}_{K^{\flat}} \subseteq \mathbb{C}^{X \times V}$ is generated by the elements $K^{\flat}_{x,v}$, $x \in X$, $v \in V$, with the inner product

$$\langle K^{\flat}_{x,v}, K^{\flat}_{y,w} \rangle = K(x, y)(v, w) =: K^{\flat}_{y,w}(x, v),$$

so that, for all $f \in \mathcal{H}_{K^{\flat}}$, we have

$$f(x, v) = \langle K^{\flat}_{x,v}, f \rangle. \tag{A.1}$$

We identify $\mathcal{H}_{K^{\flat}}$ with a subspace of the space $(V^*)^X$ of functions on X with values in the space V^* of complex-valued linear functionals on V by identifying $f \in \mathcal{H}_{K^{\flat}}$ with the function $f^*: X \to V^*$, $f^*(x) := f(x, \cdot)$. We call

$$\mathcal{H}_K := \{f^*: f \in \mathcal{H}_{K^{\flat}}\} \subseteq (V^*)^X$$

the (*vector-valued*) *reproducing kernel space associated to* K. The elements

$$K_{x,v} := (K^{\flat}_{x,v})^* \quad \text{with} \quad K_{x,v}(y) = K(y, x)(\cdot, v) \quad \text{for} \quad x, y \in X, v, w \in V,$$

then form a dense subspace of $\mathcal{H}_K$ with

$$\langle K_{x,v}, K_{y,w} \rangle = K(x, y)(v, w) \qquad . \tag{A.2}$$

and

$$\langle K_{x,v}, f \rangle = f(x)(v) \quad \text{for} \quad f \in \mathcal{H}_K, x \in X, v \in V. \tag{A.3}$$

Remark A.1.2 Equation (A.2) shows that positive definiteness of K implies the existence of a Hilbert space $\mathcal{H}$ and a map $\gamma: X \to \mathrm{Hom}(V, \mathcal{H})$, $\gamma(x)(v) := K_{x,v}$ such that

$$K(x, y)(v, w) = \langle \gamma(x)(v), \gamma(y)(w) \rangle.$$

If, conversely, such a factorization exists, then the positive definiteness follows from

$$\sum_{j,k=1}^{n} \overline{c_j} c_k K(x_j, x_k)(v_j, v_k) = \sum_{j,k=1}^{n} \overline{c_j} c_k \langle \gamma(x_j)(v_j), \gamma(x_k)(v_k) \rangle = \left\| \sum_{k=1}^{n} c_k \gamma(x_k)(v_k) \right\|^2 \geq 0.$$

Example A.1.3 If V is a complex Hilbert space, then we write $\mathrm{Sesq}(V) \subseteq \mathrm{Bil}(V)$ for the linear subspace of *sesquilinear maps*, i.e., maps which are anti-linear in the first and complex linear in the second argument. If X is a set and $K: X \times X \to B(V)$ is an operator-valued kernel, then K is positive definite if and only if the corresponding kernel

$$\tilde{K}: (X \times V) \times (X \times V) \to \mathbb{C}, \quad \tilde{K}((x, v), (y, w)) := \langle v, K(x, y)w \rangle$$

is positive definite (Definition A.1.1). Then, for each $f \in \mathscr{H}_{\tilde{K}}$, the linear functionals $f^*(x): V \to \mathbb{C}$ are continuous, hence can be identified with elements of V. Accordingly, we consider $\mathscr{H}_K$ as a space of V-valued functions (see [Nel64, Chap. 1] for more details).

Example A.1.4 Let $\mathscr{A}$ be a C^*-algebra. A linear functional $\omega \in \mathscr{A}^*$ is called *positive* if the kernel $K_\omega(A, B) := \omega(A^*B)$ on $\mathscr{A} \times \mathscr{A}$ is positive definite. Then the corresponding Hilbert space $\mathscr{H}_\omega := \mathscr{H}_{K_\omega}$ can be realized in the space $\mathscr{A}^\sharp$ of anti-linear functionals on $\mathscr{A}$. It can be obtained from the GNS representation $(\pi_\omega, \mathscr{H}_\omega, \Omega)$ [BR02, Corollary 2.3.17] by

$$\Gamma: \mathscr{H}_\omega \to \mathscr{A}^\sharp, \quad \Gamma(\xi)(A) := \langle \pi(A)\Omega, \xi \rangle$$

because $\langle \pi(A)\Omega, \pi(B)\Omega \rangle = \omega(A^*B) = K_\omega(A, B)$. Note that $\mathscr{A}$ has a natural representation on $\mathscr{A}^\sharp$ by $(A.\beta)(B) := \beta(A^*B)$ and that Γ is equivariant with respect to this representation.[1]

If $X = G$ is a group and the kernel K is invariant under right translations, then it is of the form $K(g, h) = \varphi(gh^{-1})$ for a function $\varphi: G \to \mathrm{Bil}(V)$.

Definition A.1.5 Let G be a group and let V be a real vector space. A function $\varphi: G \to \mathrm{Bil}(V)$ is said to be *positive definite* if the $\mathrm{Bil}(V)$-valued kernel $K(g, h) := \varphi(gh^{-1})$ is positive definite.

Suppose, more generally, that $(S, *)$ is an *involutive semigroup*, i.e., a semigroup S, endowed with an involutive map $s \mapsto s^*$ satisfying $(st)^* = t^*s^*$ for $s, t \in S$. A function $\varphi: S \to \mathrm{Bil}(V)$ is called *positive definite* if the kernel $K(s, t) := \varphi(st^*)$ is positive definite.

The following proposition generalizes the GNS construction to form-valued positive definite functions on groups [NÓ15b, Proposition A.4].

Proposition A.1.6 (GNS-construction for groups) *Let V be a real vector space.*

(a) *Let $\varphi: G \to \mathrm{Bil}(V)$ be a positive definite function. Then $(U_g^\varphi f)(h) := f(hg)$ defines a unitary representation of G on the reproducing kernel Hilbert space $\mathscr{H}_\varphi := \mathscr{H}_K \subseteq (V^*)^G$ with kernel $K(g, h) = \varphi(gh^{-1})$ and the range of the map*

---

[1] This realization of the Hilbert space $\mathscr{H}_\omega$ has the advantage that we can view its elements as elements of the space $\mathscr{A}^\sharp$ (see [Nel64] for many applications of this perspective). Usually, $\mathscr{H}_\omega$ is obtained as the Hilbert completion of a quotient of $\mathscr{A}$ by a left ideal which leads to a much less concrete space.

$$j: V \to \mathcal{H}_\varphi, \quad j(v)(g)(w) := \varphi(g)(w, v), \quad j(v) = K_{e,v},$$

is a cyclic subspace, i.e., $U_G^\varphi j(V)$ spans a dense subspace of $\mathcal{H}$. We then have

$$\varphi(g)(v, w) = \langle j(v), U_g^\varphi j(w) \rangle \quad \text{for} \quad g \in G, v, w, \in V. \tag{A.4}$$

(b) *If, conversely, $(U, \mathcal{H})$ is a unitary representation of G and $j: V \to \mathcal{H}$ a linear map, then*

$$\varphi: G \to \mathrm{Bil}(V), \quad \varphi(g)(v, w) := \langle j(v), U_g j(w) \rangle$$

is a $\mathrm{Bil}(V)$-valued positive definite function. If, in addition, $j(V)$ is cyclic, then $(U, \mathcal{H})$ is unitarily equivalent to $(U^\varphi, \mathcal{H}_\varphi)$.

Proof (a) For the kernel $K(g, h) := \varphi(gh^{-1})$ and $v \in V$, the right invariance of the kernel K on G implies on $\mathcal{H}_\varphi$ the existence of well-defined unitary operators U_g with

$$U_g K_{h,v} = K_{hg^{-1},v} \quad \text{for} \quad g, h \in G, v \in V.$$

In fact, (A.2) shows that

$$\langle K_{h_1 g^{-1}, v_1}, K_{h_2 g^{-1}, v_2} \rangle = K(h_1 g^{-1}, h_2 g^{-1})(v_1, v_2) = K(h_1, h_2)(v_1, v_2) = \langle K_{h_1, v_1}, K_{h_2, v_2} \rangle.$$

For $f \in \mathcal{H}_\varphi$, we then have

$$(U_g f)(h)(v) = \langle K_{h,v}, U_g f \rangle = \langle U_{g^{-1}} K_{h,v}, f \rangle = \langle K_{hg,v}, f \rangle = f(hg)(v),$$

i.e., $(U_g f)(h) = f(hg)$. Further, $j(v) = K_{e,v}$ satisfies $U_g j(v) = K_{g^{-1},v}$, which shows that $U_G j(V)$ is total in $\mathcal{H}_\varphi$. Finally we note that

$$\langle j(v), U_g j(w) \rangle = \langle K_{e,v}, K_{g^{-1},w} \rangle = K(g)(v, w) = \varphi(g)(v, w).$$

(b) The positive definiteness of φ follows with Remark A.1.2 easily from the relation $\varphi(gh^{-1})(v, w) = \langle U_g^{-1} v, U_h^{-1} w \rangle$. Since $j(V)$ is cyclic, the map $\Gamma(\xi)(g)(v) := \langle U_g^{-1} j(v), \xi \rangle$ defines an injection $\mathcal{H} \hookrightarrow (V^*)^G$ whose range is the subspace $\mathcal{H}_\varphi$ and which is equivariant with respect to the right translation representation U^φ. $\qquad\square$

Remark A.1.7 (a) If $\varphi: G \to \mathrm{Bil}(V)$ is a positive definite function, then (A.4) shows that, if $\check{V} := \overline{j(V)}$, which is the real Hilbert space defined by completing V with respect to the positive semidefinite form $\varphi(e)$, then

$$\tilde{\varphi}(g)(v, w) = \langle v, U_g w \rangle \tag{A.5}$$

defines a positive definite function

$$\tilde{\varphi} \colon G \to \mathrm{Bil}(\tilde{V}) \quad \text{with} \quad \tilde{\varphi}(g)(j(v), j(w)) = \varphi(g)(v, w) \quad \text{for} \quad v, w \in V.$$

Therefore it often suffices to consider $\mathrm{Bil}(V)$-valued positive definite functions for real Hilbert space V for which $\varphi(e)$ is a positive definite hermitian form on V whose real part is the scalar product on V. In terms of (A.4), this means that $j \colon V \to \mathcal{H}$ is an isometric embedding of the real Hilbert space V.

(b) If $\mathcal{V}$ is a real Hilbert space and j is continuous, then the adjoint operator $j^* \colon \mathcal{H} \to \mathcal{V}$ is well-defined and we obtain from (A.5) the $B(\mathcal{V})$-valued positive definite function $\varphi(g) := j^* U_g j$ which can be used to realize $\mathcal{H}$ in $\mathcal{V}^G$.

Example A.1.8 (Vector-valued GNS construction for semigroups) [Nel64, Sect. 3.1] Let $(U, \mathcal{H})$ be a representation of the unital involutive semigroup $(S, *, e)$, $\mathcal{V}$ be a Hilbert space and $j \colon \mathcal{V} \to \mathcal{H}$ be a linear map for which $U_S j(\mathcal{V})$ is total in $\mathcal{H}$. Then $\varphi(s) := j^* U_s j$ is a $B(\mathcal{V})$-valued positive definite function on S with $\varphi(e) = j^* j$ (which is $\mathbf{1}$ if and only if j is isometric) because we have the factorization

$$\varphi(st^*) = j^* U_{st^*} j = (j^* U_s)(j U_t)^*.$$

The map

$$\Phi \colon \mathcal{H} \to \mathcal{V}^S, \quad \Phi(v)(s) = j^* U_s v$$

is an S-equivariant realization of $\mathcal{H}$ as the reproducing kernel space $\mathcal{H}_\varphi \subseteq \mathcal{V}^S$, on which S acts by right translation, i.e., $(U_s^\varphi f)(t) = f(ts)$.

Conversely, let S be a unital involutive semigroup and $\varphi \colon S \to B(\mathcal{V})$ be a positive definite function. Write $\mathcal{H}_\varphi \subseteq \mathcal{V}^S$ for the corresponding reproducing kernel space with kernel $K(s, t) = \varphi(st^*)$ and $\mathcal{H}_\varphi^0$ for the dense subspace spanned by $K_{s,v} = \mathrm{ev}_s^* v, s \in S, v \in \mathcal{V}$. Then $(U_s^\varphi f)(t) := f(ts)$ defines a $*$-representation of S on $\mathcal{H}_\varphi^0$. We say that φ is *exponentially bounded* if all operators U_s^φ are bounded, so that we actually obtain a representation of S by bounded operators on $\mathcal{H}_\varphi$ (cf. [Nel64, Sect. 2.4]). Then $\mathrm{ev}_e \circ U_s^\varphi = \mathrm{ev}_s$ leads to

$$\varphi(s) = \mathrm{ev}_s \, \mathrm{ev}_e^* = \mathrm{ev}_e \, U_s^\varphi \, \mathrm{ev}_e^* \quad \text{and} \quad \varphi v = \mathrm{ev}_e^* v = K_{e,v}. \tag{A.6}$$

If $S = G$ is a group with $s^* = s^{-1}$, then φ is always exponentially bounded and the representation $(U^\varphi, \mathcal{H}_\varphi)$ is unitary.

Lemma A.1.9 *Let $(S, *, e)$ be a unital involutive semigroup and $\varphi \colon S \to B(\mathcal{V})$ be a positive definite function with $\varphi(e) = \mathbf{1}$. We write $(U^\varphi, \mathcal{H}_\varphi)$ for the representation on the corresponding reproducing kernel Hilbert space $\mathcal{H}_\varphi \subseteq \mathcal{V}^S$ by $(U^\varphi(s) f)(t) := f(ts)$. Then the inclusion $\iota \colon \mathcal{V} \to \mathcal{H}_\varphi, \iota(v)(s) := \varphi(s)v$, is surjective if and only if φ is multiplicative, i.e., a representation.*

Proof If φ is multiplicative, then $(U_s^\varphi \iota(v))(t) = \varphi(ts)v = \varphi(t)\varphi(s)v \in \iota(\mathcal{V})$. Therefore the S-cyclic subspace $\iota(\mathcal{V})$ is invariant, which implies that ι is surjective.

Suppose, conversely, that ι is surjective. Then each $f \in \mathcal{H}_\varphi$ satisfies $f(s) = \varphi(s)f(e)$. For $v \in \mathcal{V}$ and $t, s \in S$, this leads to

$$\varphi(st)v = (U_t^\varphi \iota(v))(s) = \varphi(s) \cdot (U_t^\varphi \iota(v))(e) = \varphi(s)\iota(v)(t) = \varphi(s)\varphi(t)v.$$

Therefore φ is multiplicative.   □

A.2   Integral Representations

For a realization of unitary representations associated to positive definite functions in L^2-spaces, integral representations are of crucial importance. The following result is a straight-forward generalization of Bochner's Theorem for locally compact abelian groups. Here we write $\mathrm{Sesq}^+(V) \subseteq \mathrm{Sesq}(V)$ for the convex cone of positive semidefinite forms if V is a complex linear space.

Theorem A.2.1 *Let G be a locally compact abelian group. Then a function $\varphi \colon G \to \mathrm{Sesq}(V)$ for which all functions $\varphi^{v,w} := \varphi(\cdot)(v, w), v, w \in V$, are continuous is positive definite if and only if there exists a (uniquely determined) finite $\mathrm{Sesq}^+(V)$-valued Borel measure μ on the locally compact group $\hat{G}$ such that $\hat{\mu}(g) := \int_{\hat{G}} \chi(g)\, d\mu(\chi) = \varphi(g)$ holds for every $g \in G$ pointwise on $V \times V$.*

Proof If $\varphi = \hat{\mu}$ holds for a finite $\mathrm{Sesq}^+(V)$-valued Borel measure μ on the locally compact group $\hat{G}$, then the kernel $\varphi(gh^{-1})(\xi, \eta) = \int_{\hat{G}} \chi(g)\overline{\chi(h)}\, d\mu^{\xi,\eta}(\chi)$ on $G \times V$ is positive definite because

$$\sum_{j,k=1}^n \varphi(g_j g_k^{-1})(\xi_j, \xi_k) = \sum_{j,k=1}^n \int_{\hat{G}} \chi(g_j)\overline{\chi(g_k)}\, d\mu^{\xi_j, \xi_k}(\chi)$$

$$= \sum_{j,k=1}^n \int_{\hat{G}} d\mu^{\overline{\chi(g_j)}\xi_j, \overline{\chi(g_k)}\xi_k}(\chi) = \int_{\hat{G}} d\mu^{\xi,\xi} \geq 0$$

holds for $\xi := \sum_{j=1}^n \overline{\chi(g_j)}\xi_j$ and $\mu^{\xi,\eta}(\cdot) = \mu(\cdot)(\xi, \eta)$.

Suppose, conversely, that φ is positive definite. Then Bochner's Theorem for scalar-valued positive definite functions yields for every $v \in V$ a finite positive measure μ^v on $\hat{G}$ such that

$$\varphi^{v,v}(g) = \hat{\mu}^v(g) = \int_{\hat{G}} \chi(g)\, d\mu^v(\chi).$$

By polarization, we obtain for $v, w \in V$ complex measures $\mu^{v,w} := \frac{1}{4}\sum_{k=0}^3 i^{-k} \mu^{v+i^k w}$ on $\hat{G}$ with $\varphi^{v,w} = \hat{\mu}^{v,w}$. Then the collection $(\mu^{v,w})_{v,w \in V}$ of complex measures on $\hat{G}$ defines a $\mathrm{Sesq}^+(V)$-valued measure by $\mu(\cdot)(v, w) := \mu^{v,w}$ for $v, w \in V$, and this measure satisfies $\hat{\mu} = \varphi$.   □

Remark A.2.2 Suppose that E is the spectral measure on the character group $\hat{G}$ for which the continuous unitary representation $(U, \mathscr{H})$ is represented by $U_g = \int_{\hat{G}} \chi(g) \, dE(\chi)$. Then, for $\xi \in \mathscr{H}$, the positive definite function $U^\xi(g) := \langle \xi, U_g \xi \rangle$ is the Fourier transform of the measure $E^{\xi,\xi} = \langle \xi, E(\cdot)\xi \rangle$. This establishes a close link between spectral measures and the representing measures in the preceding theorem.

The following theorem follows from [NÓ15b, Theorem B.3]:

Theorem A.2.3 (Laplace transforms and positive definite kernels) *Let E be a finite dimensional real vector space and $\mathscr{D} \subseteq E$ be a non-empty open convex subset. Let V be a Hilbert space and $\varphi \colon \mathscr{D} \to B(V)$ be such that*

(L1) *the kernel $K(x, y) = \varphi\left(\frac{x+y}{2}\right)$ is positive definite.*
(L2) *φ is weak operator continuous on every line segment in $\mathscr{D}$, i.e., all functions*
      $t \mapsto \langle v, \varphi(x + th)v \rangle, v \in V,$ are continuous on $\{t \in \mathbb{R} \colon x + th \in \mathscr{D}\}$.

Then the following assertions hold:

(i) *There exists a unique $\mathrm{Herm}^+(V)$-valued Borel measure μ on the dual space E^* such that*

$$\varphi(x) = \mathscr{L}(\mu)(x) := \int_{E^*} e^{-\lambda(x)} \, d\mu(\lambda) \quad \text{for} \quad x \in \mathscr{D}.$$

(ii) *Let $T_\mathscr{D} = \mathscr{D} + iE \subseteq E_\mathbb{C}$ be the tube domain over $\mathscr{D}$. Then the map*

$$\mathscr{F} \colon L^2(E^*, \mu; V) \to \mathcal{O}(T_\mathscr{D}, V), \quad \langle \xi, \mathscr{F}(f)(z) \rangle := \langle e_{-\bar{z}/2}\xi, f \rangle$$

is unitary onto the reproducing kernel space $\mathscr{H}_\varphi := \mathscr{H}_K$ corresponding to the kernel associated to φ. It intertwines the unitary representation

$$(U_x f)(\alpha) := e^{i\alpha(x)} f(\alpha) \quad \text{on} \quad L^2(E^*, \mu) \quad \text{and}$$
$$(\tilde{U}_x f)(z) := f(z - 2ix) \quad \text{on} \quad \mathscr{H}_\varphi.$$

(iii) *φ extends to a unique holomorphic function $\hat{\varphi}$ on the tube domain $T_\mathscr{D}$ which is positive definite in the sense that the kernel $\hat{\varphi}\left(\frac{z+\bar{w}}{2}\right)$ is positive definite.*

Corollary A.2.4 *A continuous function $\varphi \colon \mathscr{D} \to \mathbb{C}$ on an open convex subset of a finite dimensional real vector space E is positive definite if and only if there exists a positive measure μ on E^* such that $\varphi = \mathscr{L}(\mu)|_\mathscr{D}$.*

The preceding theorem generalizes in an obvious way to $\mathrm{Sesq}(V)$-valued functions, where the corresponding measure μ has values in the cone $\mathrm{Sesq}^+(V)$. One can use the same arguments as in the proof of Bochner's Theorem (Theorem A.2.1).

The following lemma sharpens the "technical lemma" in [KL82, Appendix A].
We recall the notation $\mathscr{S}_\beta = \{z \in \mathbb{C} : 0 < \mathrm{Im}\, z < \beta\}$ for horizontal strips in $\mathbb{C}$.

Lemma A.2.5 *Let $U_t = e^{itH}$ be a unitary one-parameter group on $\mathscr{H}$, E the spectral measure of H, $\xi \in \mathscr{H}$, $E^\xi := \langle \xi, E(\cdot)\xi \rangle$, $\beta > 0$ and $\varphi(t) := \langle \xi, U_t\xi \rangle = \int_{\mathbb{R}} e^{it\lambda}\, dE^\xi(\lambda)$. Then the following are equivalent:*

(i) *There exists a continuous function ψ on $\overline{\mathscr{S}_\beta}$, holomorphic on $\mathscr{S}_\beta$, such that*
     $\psi|_{\mathbb{R}} = \varphi$.
(ii) $\mathscr{L}(E^\xi)(\beta) = \int_{\mathbb{R}} e^{-\beta\lambda}\, dE^\xi(\lambda) < \infty$.
(iii) $\xi \in \mathscr{D}(e^{-\beta H/2})$.

Proof That (i) implies (ii) follows from [Ri66, p. 311]. If, conversely, (ii) is satisfied, then $\psi(z) := \mathscr{L}(E^\xi)(-iz)$ is defined on $\overline{\mathscr{S}_\beta}$, holomorphic on $\mathscr{S}_\beta$ and $\psi|_{\mathbb{R}} = \varphi$. Finally, the equivalence of (ii) and (iii) follows from the definition of the unbounded operator $e^{-\beta H/2}$ in terms of the spectral measure E. $\qquad\qquad\square$

Lemma A.2.6 (Criterion for the existence of $\mathscr{L}(\mu)(x)$) *Let $\mathscr{V}$ be a Hilbert space and μ be a finite $\mathrm{Herm}^+(\mathscr{V})$-valued Borel measure on $\mathbb{R}$, so that we can consider its Laplace transform $\mathscr{L}(\mu)$, taking values in $\mathrm{Herm}(\mathscr{V})$, whenever the integral*

$$\mathrm{tr}\left(\mathscr{L}(\mu)(x)S\right) = \int_{\mathbb{R}} e^{-\lambda x}\, d\mu^S(\lambda) \quad for \quad d\mu^S(\lambda) = \mathrm{tr}(S\, d\mu(\lambda)),$$

exists for every positive trace class operator S on $\mathscr{V}$. This is equivalent to the finiteness of the integrals $\mathscr{L}(\mu^v)(x)$ for every $v \in \mathscr{V}$, where $d\mu^v(\lambda) = \langle v, d\mu(\lambda)v \rangle$.

Proof For $x \in \mathbb{R}$, the existence of $\mathscr{L}(\mu)(x)$ implies the finiteness of the integrals $\mathscr{L}(\mu^v)(x)$ for $v \in \mathscr{V}$. Suppose, conversely, that all these integrals are finite. Then we obtain by polarization a hermitian form $\beta(v, w) := \int_{\mathbb{R}} e^{-\lambda x} \langle v, d\mu(\lambda)w \rangle$ on $\mathscr{V}$. We claim that β is continuous. As $\mathscr{V}$ is in particular a Fréchet space, it suffices to show that, for every $w \in \mathscr{V}$, the linear functional $\lambda(v) := \beta(w, v)$ is continuous [Ru73, Theorem 2.17].

The linear functionals $f_n(v) := \int_{-n}^{n} e^{-\lambda x} \langle w, d\mu(\lambda)v \rangle$ are continuous because μ is a bounded measure and the functions $e_x(\lambda) := e^{\lambda(x)}$ are bounded on bounded intervals. By the Monotone Convergence Theorem, combined with the Polarization Identity, $f_n \to f$ holds pointwise on $\mathscr{V}$, and this implies the continuity of f [Ru73, Theorem 2.8].

For a positive trace class operators $S = \sum_n \langle v_n, \cdot \rangle v_n$ with $\mathrm{tr}\, S = \sum_n \|v_n\|^2 < \infty$, we now obtain

$$\mathscr{L}(\mu^S)(x) = \sum_n \mathscr{L}(\mu^{v_n})(x) = \sum_n \beta(v_n, v_n) \le \|\beta\| \sum_n \|v_n\|^2 < \infty. \qquad\square$$

References

[AF01]   Agricola, I., Friedrich, T.: Globale Analysis, Vieweg (2001)
[AFG86] de Angelis, G., de Falco, D., Di Genova, G.: Random fields on Riemannian manifolds: a constructive approach. Commun. Math. Phys. **103**, 297–303 (1986)
[AG82]   Arsense, G., Gheondea, A.: Completing matrix contractions. J. Oper. Theory **7**, 179–189 (1982)
[AG90]   Akhiezer, D.N., Gindikin, S.G.: On Stein extensions of real symmetric spaces. Math. Ann. **286**, 1–12 (1990)
[AHH86] Albeverio, S., Høegh-Krohn, R., Holden, H.: Random fields with values in Lie groups and Higgs fields. In: Albeverio, S., Casati, G., Merlini, D. (eds.) Stochastic Processes in Classical and Quantum Systems. Lecture Notes in Physics, vol. 262, pp. 1–13
[An13]   Anderson, C.C.A.: Defining physics at imaginary time: reflection positivity for certain riemannian manifolds. Harvard University, Thesis (2013)
[Ar50]   Aronszajn, N.: Theory of reproducing kernels. Trans. Am. Math. Soc. **68**, 337–404 (1950)
[AS80]   Alonso, A., Simon, B.: The Birman-Krein-Vishik theory of selfadjoint extensions of semibounded operators. J. Oper. Theory **4**, 251–270 (1980)
[vB88]   van den Ban, E.: The principal series for a reductive symmetric space, I. H-fixed distribution vectors. Ann. Sci. Ec. Norm. Super. **21**, 359–412 (1988)
[Ba78]   Bauer, H.: Wahrscheinlichkeitstheorie und Grundzüge der Maßtheorie. Walter de Gruyter, Berlin (1978)
[Ba96]   Bauer, H.: Probability Theory. Studies in Mathematics, vol. 23. Walter de Gruyter, Berlin (1996). Sect. 36 on kernels
[BGN17] Beltita, D., Neeb, K.-H., Grundling, H.: Covariant representations for singular actions on C^*-algebras, 76pp. arXiv:math.OA:1708.01028
[BJM16] Barata, J.C.A., Jäkel, C.D., Mund, J.: Interacting quantum fields on de Sitter space. arXiv:math-ph:1607.02265
[Bo92]   Borchers, H.-J.: The CPT-theorem in two-dimensional theories of local observables. Commun. Math. Phys. **143**, 315–332 (1992)
[BLS11] Buchholz, D., Lechner, G., Summers, S.J.: Warped convolutions, Rieffel deformations and the construction of quantum field theories. Commun. Math. Phys. **304**(1), 95–123 (2011)
[BR02]   Bratteli, O., Robinson, D.W.: Operator Algebras and Quantum Statistical Mechanics II. Texts and Monographs in Physics, 2nd edn. Springer, Berlin (1996)
[BR96]   Bratteli, O., Robinson, D.W.: Operator Algebras and Quantum Statistical Mechanics I. Texts and Monographs in Physics, 2nd edn. Springer, Berlin (2002)

© The Author(s) 2018
K.-H. Neeb and G. Ólafsson, *Reflection Positivity*, SpringerBriefs in Mathematical Physics, https://doi.org/10.1007/978-3-319-94755-6

[DG13]   Dereziński, J., Gérard, C.: Mathematics of Quantization and Quantum Fields. Cambridge Monographs on Mathematical Physics. Cambridge University Press, Cambridge (2013)

[Di04]   Dimock, J.: Markov quantum fields on a manifold. Rev. Math. Phys. **16**(2), 243–255 (2004)

[vD09]   van Dijk, G.: Introduction to Harmonic Analysis and Generalized Gelfand Pairs. Studies in Mathematics. de Gruyter, Berlin (2009)

[DM78]   Dixmier, J., Malliavin, P.: Factorisations de fonctions et de vecteurs indéfiniment différentiables. Bull. Soc. Math. 2e série **102**, 305–330 (1978)

[EN00]   Engel, K.-J., Nagel, R.: One-parameter Semigroups for Linear Evolution Equations. Graduate Texts in Mathematics, vol. 194. Springer, New York (2000)

[Fa00]   Fabec, R.C.: Fundamentals of Infinite Dimensional Representation Theory. Chapman & Hall/CRC, Boca Raton (2000)

[Fa08]   Faraut, J.: Analysis on Lie Groups. An Introduction. Cambridge Studies in Advanced Mathematics, vol. 110. Cambridge University Press, Cambridge (2008)

[FK94]   Faraut, J., Korányi, A.: Analysis on Symmetric Cones. Clarendon Press, Oxford (1994)

[FILS78]   Fröhlich, J., Israel, R., Lieb, E.H., Simon, B.: Phase transitions and reflection positivity. I. General theory and long range lattice models. Commun. Math. Phys. **62**(1), 1–34 (1978)

[FL10]   Frank, R.L., Lieb, E.H.: Inversion positivity and the sharp Hardy-Littlewood-Sobolev inequality. Calc. Var. Partial. Differ. Equ. **39**, 85–99 (2010)

[FL11]   Frank, R.L., Lieb, E.H.: Spherical reflection positivity and the Hardy-Littlewood-Sobolev inequality. Contemp. Math. **545**, 89–102 (2011)

[FNO18]   Frahm, J., Neeb, K.-H., Ólafsson, G.: Vector-valued reflection positivity on the sphere (in preparation)

[Fo95]   Folland, G.B.: A Course in Abstract Harmonic Analysis. CRC Press, Boca Raton (1995)

[FÓØ18]   Frahm, J., Ólafsson, G., Ørsted, B.: The Berezin form on symmetric R-spaces and reflection positivity. In: Grabowska, K., Grabowski, J., Fialowski, A., Neeb, K.-H. (eds.) 50th Seminar Sophus Lie, vol. 113, pp. 135–168. Banach Center Publications (2018)

[FOS83]   Fröhlich, J., Osterwalder, K., Seiler, E.: On virtual representations of symmetric spaces and their analytic continuation. Ann. Math. **118**, 461–489 (1983)

[Fro80]   Fröhlich, J.: Unbounded, symmetric semigroups on a separable Hilbert space are essentially selfadjoint. Adv. Appl. Math. **1**(3), 237–256 (1980)

[Fro11]   Fröhlich, J.: Phase transitions and continuous symmetry breaking. Lecture Notes, Vienna (2011)

[GHL87]   Gallot, S., Hulin, D., Lafontaine, J.: Riemannian Geometry. Springer, Universitext (1987)

[GJ81]   Glimm, J., Jaffe, A.: Quantum Physics-A Functional Integral Point of View. Springer, New York (1981)

[HH17]   Helleland, C., Hervik, S.: Wick rotations and real GIT. arxiv:math.DG:1703.04576v1

[Hid80]   Hida, T.: Brownian Motion. Applications of Mathematics, vol. 11. Springer, Berlin (1980)

[Hi89]   Hilgert, J.: A note on Howe's oscillator semigroup. Ann. Inst. Fourier **39**, 663–688 (1989)

[HN12]   Hilgert, J., Neeb, K.-H.: Lie Semigroups and Their Applications. Lecture Notes in Mathematics, vol. 1552. Springer, Berlin (1993)

[HN93]   Hilgert, J., Neeb, K.-H.: Structure and Geometry of Lie Groups. Springer, New York (2012)

[HÓ97]   Hilgert, J., Ólafsson, G.: Causal Symmetric Spaces, Geometry and Harmonic Analysis. Perspectives in Mathematics, vol. 18. Academic Press, San Diego (1997)

[How88]   Howe, R.: The oscillator semigroup. In: Wells, R.O. (ed.) The Mathematical Heritage of Hermann Weyl. Proceedings of Symposia in Pure Mathematics, vol. 48. American Mathematical Society, Providence (1988)

[Ja08]   Jaffe, A.: Quantum theory and relativity. In: Doran, R.S., Moore, C.C., Zimmer, R.J. (eds.) Group Representations, Ergodic Theory, and Mathematical Physics: A Tribute to George W. Mackey. Contemporary in Mathematics, vol. 449. American Mathematical Society, Providence (2008)

[JR07]   Jaffe, A., Ritter, G.: Quantum field theory on curved backgrounds. I. The euclidean functional integral. Commun. Math. Phys. **270**, 545–572 (2007)

[Jo86]   Jorgensen, P.E.T.: Analytic continuation of local representations of Lie groups. Pac. J. Math. **125**(2), 397–408 (1986)

[Jo87]   Jorgensen, P.E.T.: Analytic continuation of local representations of symmetric spaces. J. Funct. Anal. **70**, 304–322 (1987)

[Jo02]   Jorgensen, P.E.T.: Diagonalizing operators with reflection symmetry. J. Funct. Anal. **190**, 93–132 (2002)

[JN16]   Jorgensen, P.E.T., Neeb, K.-H., Ólafsson, G.: Reflection positive stochastic processes indexed by Lie groups. SIGMA **12**, 058, 49 pp. (2016). arXiv:math-ph:1510.07445

[JNO18] Jorgensen, P.E.T., Neeb, K.-H., Ólafsson, G.: Reflection positivity on real intervals. Semigroup Forum **96**, 31–48 (2018)

[JN15]   Jorgensen, P.E.T., Niedzialomski, R.: Extension of positive definite functions. J. Math. Anal. Appl. **422**, 712–740 (2015)

[JÓl98] Jorgensen, P.E.T., Ólafsson, G.: Unitary representations of Lie groups with reflection symmetry. J. Funct. Anal. **158**, 26–88 (1998)

[JÓl00] Jorgensen, P.E.T., Ólafsson, G.: Unitary representations and Osterwalder–Schrader duality. In: Doran R.S., Varadarajan V.S. (eds.) The Mathematical Legacy of Harish–Chandra. Proceedings of Symposia in Pure Mathematics, vol. 68, pp. 333–401. American Mathematical Society (2000)

[JPT15] Jorgensen, P.E.T., Pedersen, S., Tian, F.: Extensions of Positive Definite Functions: Applications and Their Harmonic Analysis. Lecture Notes in Mathematics, vol. 2160. Springer, Berlin (2016)

[JT17]   Jorgensen, P.E.T., Tian, F.: Reflection positivity and spectral theory 15 May 2017. arXiv:1705.05262v1 [math.FA]

[Ka85]   Kay, B.S.: A uniqueness result for quasi-free KMS states. Helv. Phys. Acta **58**(6), 1017–1029 (1985)

[Ka85b] Kay, B.S.: Purification of KMS states. Helv. Phys. Acta **58**(6), 1030–1040 (1985)

[Kl77]   Klein, A.: Gaussian OS-positive processes. Z. Wahrscheinlichkeitstheorie und Verw. Gebiete **40**(2), 115–124 (1977)

[Kl78]   Klein, A.: The semigroup characterization of Osterwalder-Schrader path spaces and the construction of euclidean fields. J. Funct. Anal. **27**, 277–291 (1978)

[KL75]   Klein, A., Landau, L.: Singular perturbation of positivity preserving semigroups via path space techniques. J. Funct. Anal. **20**(1), 44–82 (1975)

[KL81]   Klein, A., Landau, L.: Periodic Gaussian Osterwalder-Schrader positive processes and the two-sided Markov property on the circle. Pac. J. Math. **94**(2), 341–367 (1981)

[KL81b] Klein, A., Landau, L.: Stochastic processes associated with KMS states. J. Funct. Anal. **42**(3), 368–428 (1981)

[KL82]   Klein, A., Landau, L.: From the Euclidean group to the Poincaré group via Osterwalder–Schrader positivity. Commun. Math. Phys. **87**, 469–484 (1982/1983)

[KO08]   Krötz, B., Opdam, E.: Analysis on the crown domain. Geom. Funct. Anal. **18**(4), 1326–1421 (2008)

[KS04]   Krötz, B., Stanton, R.J.: Holomorphic extensions of representations. I. Automorphic functions. Ann. Math. **159**, 641–724 (2004)

[KS05]   Krötz, B., Stanton, R.J.: Holomorphic extensions of representations II. Geometry and harmonic analysis. Geom. Funct. Anal. **15**(1), 190–245 (2005)

[LP64]   Lax, P.D., Phillips, R.S.: Scattering theory. Bull. Am. Math. Soc. **70**, 130–142 (1964)

[LP67]   Lax, P.D., Phillips, R.S.: Scattering Theory. Pure and Applied Mathematics, vol. 26. Academic Press, New York (1967)

[LP81]   Lax, P.D., Phillips, R.S.: The translation representation theorem. Integral Equ. Oper.
            Theory **4**, 416–421 (1981)
[La94]   Lawson, J.D.: Polar and Olshanskii decompositions. J. reine und angew. Math. **448**,
            191–219 (1994)
[LL14]   Lechner, G., Longo, R.: Localization in nets of standard spaces. Commun. Math. Phys.
            336(1), 27–61 (2015)
[LM75]   Lüscher, M., Mack, G.: Global conformal invariance and quantum field theory. Commun.
            Math. Phys. **41**, 203–234 (1975)
[Lo08]   Longo, R.: Real Hilbert subspaces, modular theory, SL(2, R) and CFT. Von Neumann
            Algebras in Sibiu. Theta Foundation International Book Series of Mathematical Texts,
            vol. 10, pp. 33–91. Theta, Bucharest (2008)
[LW11]   Longo, R., Witten, E.: An algebraic construction of boundary quantum field theory.
            Commun. Math. Phys. **303**(1), 213–232 (2011)
[Luk70]   Lukacs, E.: Characteristic Functions. Griffin, London (1970)
[MN12]   Merigon, S., Neeb, K.-H.: Analytic extension techniques for unitary representations of
            Banach-Lie groups. Int. Math. Res. Not. **18**, 4260–4300 (2012)
[MNO15]   Merigon, S., Neeb, K.-H., Ólafsson, G.: Integrability of unitary representations on repro-
            ducing kernel spaces. Represent. Theory **19**, 24–55 (2015)
[Ne98]   Neeb, K.-H.: Operator valued positive definite kernels on tubes. Monatshefte für Math.
            126, 125–160 (1998)
[Ne18b]   Neeb, K.-H.: On the geometry of standard subspaces. In: Christensen, J.G., Dann, S.,
            Dawson, M. (eds.) Representation Theory, Symmetric Spaces, and Integral Geome-
            try. Contemporary Mathematics. American Mathematical Society, Providence (2588).
            arXiv:math.OA:1707.05506
[Ne00]   Neeb, K.-H.: Holomorphy and Convexity in Lie Theory. Expositions in Mathematics,
            vol. 28. de Gruyter Verlag, Berlin (2000)
[Nel64]   Nelson, E.: Feynman integrals and the Schrödinger equation. J. Math. Phys. **5**, 332–343
            (1964)
[Ne18]   Neeb, K.-H.: Reflection positive symmetric operators (in preparation)
[NÓ14]   Neeb, K.-H., Ólafsson, G.: Reflection positivity and conformal symmetry. J. Funct.
            Anal. **266**, 2174–2224 (2014)
[NÓ15a]   Neeb, K.-H., Ólafsson, G.: Reflection positive one-parameter groups and dilations. Com-
            plex Anal. Oper. Theory **9**(3), 653–721 (2015)
[NÓ15b]   Neeb, K.-H., Ólafsson, G.: Reflection positivity for the circle group. In: Proceedings of
            the 30th International Colloquium on Group Theoretical Methods; J. Phys. Conf. Ser.
            597, 012004 (2015). arXiv:math.RT.1411.2439
[NÓ16]   Neeb, K.-H., Ólafsson, G.: KMS conditions, standard real subspaces and reflection
            positivity on the circle group (submitted). arXiv:math-ph:1611.00080
[NÓ17]   Neeb, K.-H., Ólafsson, G.: Antiunitary representations and modular theory. In:
            Grabowska, K., Grabowski, J., Fialowski, A., Neeb, K.-H. (eds.) 50th Sem-
            inar Sophus Lie, vol. 113, pp. 291–363 (2018). Banach Center Publications.
            arXiv:math-RT:1704.01336
[NÓ18]   Neeb, K.-H., Ólafsson, G.: Reflection positivity on spheres (in preparation)
[Nel69]   Nelson, E.: Analytical vectors. Ann. Math. **70**, 572–615 (1969)
[Nel73]   Nelson, E.: Construction of quantum fields from Markoff fields. J. Funct. Anal. **12**,
            97–112 (1973)
[Ól00]   Ólafsson, G.: Analytic continuation in representation theory and harmonic analysis.
            Séminaire et Congrès **4**, 201–233 (2000)
[ÓP13]   Ólafsson, G., Pasquale, A.: Ramanujan's Master Theorem for the hypergeometric
            Fourier Transform associated with root systems. J. Fourier Anal. **19**, 1150–1183 (2013)
[OS73]   Osterwalder, K., Schrader, R.: Axioms for Euclidean green's functions. I. Commun.
            Math. Phys. **31**, 83–112 (1973)
[OS75]   Osterwalder, K., Schrader, R.: Axioms for Euclidean Green's functions. II. Commun.
            Math. Phys. **42**, 281–305 (1975)

[Po72]   Poulsen, N.S.: On C^∞-vectors and intertwining bilinear forms for representations of
          Lie groups. J. Funct. Anal. **9**, 87–120 (1972)
[RS75]   Reed, S., Simon, B.: Methods of Mathematical Physics II: Fourier Analysis. Self-
          adjointness. Academic Press, New York (1975)
[Ri66]   Richter, H.: Wahrscheinlichkeitstheorie, vol. 86, 2nd edn. Springer, Grundlehren (1966)
[RvD77]   Rieffel, M.A., van Daele, A.: A bounded operator approach to Tomita-Takesaki Theory.
          Pac. J. Math. **69**(1), 187–220 (1977)
[Ru73]   Rudin, W.: Functional Analysis. McGraw Hill, New York (1973)
[SSV10]   Schilling, R., Song, R., Vondracek, Z.: Bernstein Functions. Studies in Mathematics. de
          Gruyter, Berlin (2010)
[Sch99]   Schlingemann, D.: From euclidean field theory to quantum field theory. Rev. Math.
          Phys. **11**, 1151–1178 (1999)
[Sch73]   Schwartz, L.: Théorie des distributions, 2nd edn. Hermann, Paris (1973)
[Sim05]   Simon, B.: Functional Integration and Quantum Physics, 2nd edn. AMS Chelsea Pub-
          lishing, Providence (2005)
[Sin61]   Dynamical systems with countable Lebesgue spectrum: Sinai, Ya.G. I. I. Izv. Akad.
          Nauk. USSR **25**, 899–924 (1961)
[Str83]   Strichartz, R.S.: Analysis of the Laplacian on the complete Riemannian manifold. J.
          Funct. Anal. **52**(1), 48–79 (1983)
[SzN10]   Sz.-Nagy, B., Foias, C., Bercovici, H., Kérchy, L.: Harmonic Analysis of Operators on
          Hilbert Space (Universitext), 2nd edn. Springer, New York (2010)
[Tr67]   Treves, F.: Topological Vector Spaces, Distributions, and Kernels. Academic Press, New
          York (1967)
[Us12]   Usui, K.: A note on reflection positivity and the Umezawa–Kamefuchi–Källén-Lehmann
          representation of two point correlation functions. arXiv:hep-lat:1201.3415v4
[VW90]   Vogan Jr., D.A., Wallach, N.R.: Intertwining operators for real reductive groups. Adv.
          Math. **82**, 203–243 (1990)
[Wa72]   Warner, G.: Harmonic Analysis on Semisimple Lie Groups I. Springer, Berlin (1972)
[WW63]   Whittaker, E.T., Watson, G.N.: A Course of Modern Analysis, 4th edn. Cambridge
          University Press, London (1963)
[Wi34]   Widder, D.V.: Necessary and sufficient conditions for the representation of a function
          by a doubly infinite Laplace integral. Bull. Am. Math. Soc. **40**(4), 321–326 (1934)
[Wi46]   Widder, D.V.: The Laplace Transform. Princeton University Press, Princeton (1946)
[Za17]   Zahariev, S.: On scaling limits in euclidean quantum field theory.
          arxiv:math-ph:1701.05569v1

Index

Symbols

(G, S, τ)-measure space, 114
(G, S, τ)-measure space, Markov, 114
(G, τ)-measure space, 113
(G, τ)-measure space, reflection positive, 114
(G, τ)-probability space, 113
C^*-dynamical system, 55
$C^{-\infty}(X)$: space of distributions, 14
$C_\lambda^{-\infty}$, 88
C_λ^{∞}, 88
G^τ, 2
$G_\tau = G \rtimes \{\mathrm{id}_G, \tau\}$, 24
$\mathscr{E}_0$, 11
θ-positive subspace, 9
θ-positive subspace of Hilbert space, 9
$\mathfrak{B}(X)$, Borel subsets of top. space X, 48

C

Conjugation
  on Hilbert space, 6, 51
Convolution product, 80
Convolution semigroup, 121
Crown domain, 97

D

De Sitter space, dS^n, 26, 100
Differential equation
  hypergeometric, 98
Distribution
  positive definite, 83
  reflection positive, 85, 86
Distribution kernel
  β-compatible, 72, 75
  positive definite, 14

reflection positive, 14
Distribution vector, 82
  cyclic, 83
  reflection positive, 86
Dual symmetric Lie algebra, 1

E

Euclidean realization, 39, 62
Euclidean realization: of contraction representation, 28
Euclidean realization: of unitary representation of G^c, 28

F

Feynman–Kac–Nelson formula, 120
Forward light cone, 8
Fourier transform
  of a function, 8
  of a measure, 8
Function
  completely monotone, 36
  form-valued
    holomorphic, 52
    pointwise continuous, 52
  reflection positive, 37
  spherical, 92
  τ-positive definite, 73

H

Hardy space, 44
Hilbert space
  reflection positive, 10
  reproducing kernel, 124

© The Author(s) 2018
K.-H. Neeb and G. Ólafsson, *Reflection Positivity*, SpringerBriefs in Mathematical Physics, https://doi.org/10.1007/978-3-319-94755-6

K

Kernel
   Markov, 119
   on measurable spaces, 119
   positive definite, 123
      operator-valued, 124
   reproducing, 123
KMS condition, 5
   positive definite function, 51, 52
KMS state, 56

L

Laplace–Beltrami operator, 16
Laplacian, on $\mathbb{R}^n$, 122
Lie algebra
   dual symmetric, 24
   symmetric, 1, 24
   symmetric dual, 1
Lie derivative, 71
Lie group
   symmetric, 1, 24

M

Markov condition, 11
Markov property, 114
Measure
   Haar–Wiener, 122
   Lebesgue–Wiener, 122
   tame Borel, 105
Minimal unitary dilation, 42
Modular function, 80
Modular objects, 6, 51, 53
Modular relation, 53

O

OS transform
   of a representation, 28
OS transform: of an operator, 22
Osterwalder–Schrader (OS) transform, 2, 22
Outgoing
   realization, 48
   subspace, 47

P

Partial order $\prec_S$, 118
Path group, 120
Positive definite
   distribution kernel, 14
   function
      on group, 125

      on involutive semigroup, 125
Positive functional, 125
Positive semigroup structure, 114
   associated, 118
   standard, 115

Q

Quasi-regular representation, on $L^2(G/H)$, 83

R

Reflection
   dissecting, 16
   of Riemannian manifold, 16
Reflection positive
   cyclic representation, 32
   distribution, 85
   distribution kernel, 14
   function, 37
   function, w.r.t. subsemigroup S, 30
   function, w.r.t. subset G_+, 29
   Hilbert space, 2
      of Markov type, 11
   kernel, 12
   operator, 18
   pre-Hilbert space, 29
   unitary one-parameter group, 38
   V-cyclic representation, 32
Representation
   complementary series, 91
   distribution cyclic, 86
   exponentially bounded, 127
   highest weight, 95
   infinitesimally reflection positive, 28
   negative energy, 95
   positive energy, 109
   reflection positive
      w.r.t. subsemigroup S, 3, 27
      w.r.t. subset G_+, 3, 27

S

Semigroup
   involutive, 125
   of Markov kernels, 119
Sesquilinear maps on V, Sesq(V), 125
Shilov boundary, 100
Smooth right action, 72
Smooth vector, 81
Standard subspace, 53
Stochastic process, 117
   distribution of, 117

full, 117
  state space of, 117
  stationary, 117
  τ-invariant, 117
Strict contraction, 58
Subsemigroup
  symmetric, 3, 25
Symmetric semigroup, 25

T
Theorem
  Bochner, 128
  Characterization of reflection positive
    functions on interval, 38
  Fröhlich's Selfadjointness, 70
  Geometric Fröhlich, 71
  Geometric Fröhlich for distributions, 75
  GNS construction for reflection positive
    functions, 31
  Hausdorff–Bernstein–Widder, 36
  KMS Characterization, 54

Laplace transforms and positive definite
  kernels, 129
Lax–Phillips Representation, 47
Realization in spaces of distributions, 84
Realization Theorem for unitary one-
  parameter groups, 62
Reconstruction, for positive semigroup
  structures, 118
Reflection positive extension, 59
Widder, 36
Time translation semigroup, 104
Time-zero subspace, 107
Total subsets of Hilbert space, 7
Two-sided path space, 121

V
Vector field
  D-skew-symmetric, 75
  D-symmetric, 75
  K-skew-symmetric, 71
  K-symmetric, 71

Printed in the United States
By Bookmasters